No Tomorrow

Philip Machanick

Second edition: minor corrections and reformatting for US printing (fifth printing, further minor corrections, July 2011).

National Library of Australia Cataloguing-in-Publication entry:

Author:	Machanick, Philip, 1957-
Title:	No tomorrow / Philip Machanick
Edition:	2nd ed.
Publisher:	Taringa, Qld : RAMpage Research, 2008.
ISBN:	978-0-9804510-1-6 (pbk.)
Subjects:	Climatic changes–Fiction.
	Self-perception–Fiction.
	Documentary films–Fiction.
Dewey Number:	A823.4

To the future ... and those who want one.

Contents

1   Prentice Productions

B ILL PRENTICE was a busy man – time was money, even if he was his own boss. Queensland may be the "sunshine state", the most laid back part of Australia, but he wasn't your average Queenslander. Neither, he noted, was this kid, who seemed to be in a hurry.

"Hi, I'm Martin. Martin Truscott." The visitor thrust his hand into Prentice's space. Prentice inspected it for a few seconds, and took it, as if he already had more than he needed.

"So, what do you want?"

"I have this program idea." Truscott leant forward on his chair.

"Go on." There was a hint of impatience. Bill Prentice didn't get to be a TV producer by wasting time on crap. He eased his gut into a more comfortable position.

"You remember that *Global Warming Swindle* doco – the one that caused all the controversy..."

"Of course," Prentice snapped.

"See, Durkin stuffed it up. He started out with the position that IPCC scientists are a bunch of liars, filmed a few clips to back that position up, and stitched the thing together to support his case. So: what if the science *really* is flawed, but it takes digging a bit deeper than that? Is the IPCC ... ?"

"IPCC?" Prentice grunted.

"Intergovernmental Panel on Climate Change."

"I know what the IPCC is. But who cares? Durkin had his five minutes of fame. Who's going to buy a similar movie? The commercials won't touch it. ABC did it already. SBS – not likely: they are going for market share. And after Al Gore and the IPCC won the Nobel, is there still a story in this?"

"*Swindle* attracted a big audience. The denial side feel their case wasn't really put – and they will only be fired up even more by the Nobel. The other side is still steamed over the fact that *Swindle* was aired at all. And if I do a good job of this, it's not just the Australian market. *Swindle* went international, and it was a piss-poor..."

"OK, OK. What's the bottom line?" He paused to recollect the name. "Truscott." Prentice took pride in the details. Though there wasn't much remarkable about the encounter, he always took the trouble to remember a name; speaking of which: "And what do you already have to your name?"

"Right. I already made a couple of segments for *Catalyst*. Why multicore computing is the next big thing. Why melatonin isn't a wonder drug, how computers in schools are killing science teaching," he reached into his bag, "how low levels of dioxins turn into high levels through bioaccumulation ..." and produced some DVDs. "And I have a draft budget." His hand plunged into the bag again and retrieved a single A4 page.

Prentice ignored the DVDs and scanned the budget. "What the f—? What kind of movie can you make for this?"

His brow creased as he looked at the detail. "Budget air fares for one – where can you get such cheap flights?"

"If you aren't fussy about exactly when you go..."

Prentice waved the page at his visitor. "And what about your film crew?"

"This is for a pilot. I was thinking to use a handcam to get the main issues in, then we can look at costing the whole thing if it's a go."

"Accommodation thirty bucks a night. Where the hell can you stay for that?"

"Hostels."

"For crissake. OK. I can advance you this much. But if your pilot is no good, not a cent more. You must be effing crazy."

"Not crazy. Hungry." Truscott reached forward to shake the bigger man's hand.

"Just as well. Making docos isn't therapy." He held Truscott in a tight gaze. "It's good being focused, but you could do with turning the intensity down a notch. Now get the hell out of here, before I change my mind."

Truscott definitely didn't present a picture of affluence – a man on the bones of his backside. As he got up and turned for the door, Prentice almost felt that the kid would disappear if seen side-on. He glanced down at his gut thinking he could afford to go hungry a bit. How could someone get like that? This is a country where people get fat on the dole. Then again, in his own early days, he'd been pretty intense.

As the door closed, Prentice sighed. "Should be able to chalk this one up to some government subsidy or something. Jesus, I forgot what it was like to be like that."

2 Close to Home

IT WAS A SUNNY DAY on the Sunshine Coast, aptly named for once. It was a marketing trick to call the part of the Queensland coast with some of the coolest weather the "Sunshine" Coast. Martin was in a rental car, taking a leisurely drive to Noosa, to meet one of the most prolific writers – at least in the press – on the errors of climate science.

He mulled over Prentice's parting words. *Therapy*? Was there something wrong with him? Why was he so driven – always chasing after the elusive truth? What had him so fired up over the climate change thing? Prentice was right: it ought to be old news. But he had a nagging feeling that the story wasn't over yet. He was so intense – a loner most of his life. Why had his mum disappeared from his life? Why did his dad cut him off from family? ...Julie ...could that be it? With her holier-than-thou green attitude? Then suddenly dropping him without reason, after taking a job in what she would have described as a sell-out industry, advertising?

After all these years, thinking about Julie still hurt. He shook his head, and focused on the problem at hand. Maybe it was therapy of a kind.

"Let's see . . . " he composed his thoughts, while watching the unaccustomed scenery glide by.

"The big-picture argument is that the climate scientists are ignoring big areas of science, that they are ignoring the paleoclimatic record where CO_2 increased only after warming, they are only doing statistical fits to the data, not looking at the physics . . .

"This looks like the off-ramp." His attention switched to the *Refidex* book on the seat next to him, the indispensable local book of maps, half the thickness of a telephone directory. As he hit the first traffic lights, he took a closer look. A few more then left, right. He may have had a good memory for science trivia, but navigation wasn't his strong suit. A few wrong turns later – the scenic route, he thought wryly – he was in the right street, and looking for number forty-two. "The answer to the fundamental question of life, the universe and everything. . . Let's see if McCarthy lives up to expectation."

He parked and surveyed the house: a traditional Queenslander, on stilts (though with a concession to modernity, the stilts were made of concrete, not wood). "I wonder how the termites like that." The traditional construction was wooden stilts with a metal saucer on top. Termites don't like to be exposed to air and can't eat their way through metal, so you knew your house was under attack when you saw a mud tunnel constructed by termites to bridge over the metal. With the rest of the house mostly wood, without this early warning system, termites could find a path into the house and you wouldn't know until you crashed through a weakened floorboard, or tried to hang a picture, and hit empty space.

The house was surrounded by vegetation, an indication of the relatively high rainfall in the area. He approached the front door, checking his watch – despite the unintended scenic detour, he was about on time. He knocked. After a pause, he heard footsteps, and a greying man with salt and pepper beard, on the edge of portly but reasonably fit-looking, opened the door. McCarthy blinked in the bright sunshine. "Where's your crew?"

"Martin Truscott – sorry, it's just me. I am doing the pilot on a tight budget." He held out his hand.

McCarthy did nothing for a moment, then realized there were social niceties to dispose of. "Terribly sorry, Kev Mc-Carthy. Do come in." A brief handshake, and Martin was across the portal.

They were sitting in overstuffed chairs in a room full of marine specimens, the untidy detritus of a busy life. Martin was taking notes.

"So you worked in marine paleontology until you retired?"

"That's right."

"I found a fair number of papers you've published in that field: you have an impressive track record."

"Thanks."

"Why did you start taking an interest in climate change?"

"Climate change isn't something new – it happens all the time. In the fossil record, there are enormous variations from the climate we have today. Ice ages, extremely warm periods, everything between. It's a natural process, not something invented in the latter part of the twentieth century."

"I think everyone understands that. The big question is

whether anthropogenic climate change is real and if so, what to do about it."

"Obviously. Now if you look at the paleoclimatic record, you see that what we are seeing today is nothing unusual. Climate change on this scale has happened before, so that makes me suspicious of arguments that humans must be driving it. Let's look at some inconsistencies.

"If things were getting warmer, we would expect a continuing sequence of warmer years, yet 1998 was a peak year, and it hasn't become warmer since. If CO_2 is a major driver of climate change, why does the ice core record consistently show CO_2 increases *following* warming?"

He paused to take a breath, and Martin took this as his cue to ask some questions.

"That 1998 thing: wasn't 1998 an unusually warm year for other reasons like an El Niño event? And I've seen somewhere that we've now at least matched that record without an El Niño."

"I am yet to be convinced. The CO_2 argument doesn't match anything in the past, so it has to be rubbish."

Martin wasn't convinced. "But why? Dinosaurs didn't have coal-fired power plants. Nothing in nature was pumping billions of tonnes of CO_2 into the atmosphere on a consistent basis. Aren't we dealing with something new?"

"Have you seen how CO_2 levels are measured? We are talking about parts per *million*." McCarthy waived his arms animatedly. "Billions of tons sounds impressive, but those numbers are tiny compared with the total size of the atmosphere. That's about five times ten to the eighteen kilograms – you think a billion tonnes sounds like a lot: that's five million billion tonnes."

"I see." Martin paused. "But surely one can do a simple calculation of the physics here: so much CO_2 in the air means so much greenhouse effect? Isn't that the important thing rather than how big a fraction CO_2 is of the total?"

"Have you done that calculation? It's not so simple. CO_2 is water soluble, and there is a continuous exchange of CO_2 between the sea and the atmosphere. I am very suspicious of people claiming to model something as complex as the climate using statistical approaches. This is a vastly complex physical system. You have to go to the observed data, model it, then check against other observations. You need physical models – models taking into account the physics – and the overall system is too complex to do this convincingly."

"That gives me something to work on." Martin nodded. "I've also seen something you wrote about the urban heat island effect."

"Right. UHI is not taken into account adequately. A lot of weather stations are in urban areas, which are naturally warmer than rural areas. Asphalt absorbs heat. Air conditioners vent into the air ... "

Martin stopped him. "I saw your article covering all that – but surely the climate modelers should be aware of that, and allow for it?"

"You would think so, wouldn't you? But they have a point to prove – and if they don't make the trend sound catastrophic enough, how are they going to justify future funding?"

Martin stopped talking for a few minutes after McCarthy finished, to consult his notes.

"What about your own work? I found a lot of your paleobiology papers, but only one climate change paper that you've published, aside from press articles. Are you working on a comprehensive model of the climate?"

"Of course not. I'm retired. Where would I get the resources to do that? And in case you were about to ask, no, I have no funding from the fossil fuel industry."

"I understand – it's just that your critique of the IPCC implies that their modeling is deficient, so I was hoping to find something better."

"Why don't you talk to people like Schoor at Harvard? He's doing real work, not just looking for holes in the science as I am – not that my work is less legitimate for that."

"Schoor." Martin contemplated. "He's the one who's been arguing that it's all solar effects, isn't it?"

"Good lad." McCarthy nodded. "You've been doing your homework. There's also Hayes at MIT..."

"The clouds guy..."

"Great, you know more than all the other journos who've visited me put together. I'll see if I can get you time with both of them. Not that I control their schedules, but it may help if I put in a good word."

"Thanks. I think I would like to come back to you for the interview after going overseas, if you don't mind. I would like to have a more complete picture before I start putting things on the record."

"I appreciate your thoroughness. Most journos show up for a photo shoot, record a few sound bites then make something up – if they didn't make it up in advance. I have to write my own articles to get the message straight."

"And you've written a lot of those. Thanks for your time."

McCarthy led him to the door. "Let me know how your trip goes. I can always handle questions by email."

3 Sydney

THE UNIVERSITY of New South Wales – UNSW – in Randwick, Sydney is an urban campus – even if it has some nice grassed spaces in the interior to give an impression of elbow room. Its buildings have an old city campus feel – a smell of dust too deep in the crevices ever to clean out.

Dr Gene Wilkinson was starting to work on a research grant – for a change, way before the deadline – so an interruption wasn't entirely welcome. And after so much misreporting in the press, he had major misgivings. Still, Truscott sounded OK. He made a big thing of his science background, even if it was not in anything related to climate science. And he had done a few segments on *Catalyst*, one of the better science programs. Wilkinson vaguely remembered one about dioxins, a good summary of the quantitative science on bioaccumulation, not the usual fuzzy dumbed down stuff.

So when Truscott knocked on his door, he managed a good approximation to pleasantness.

"Come in, come in. What exactly is your angle?"

Truscott found a chair not totally covered with papers to sit on.

"You remember that *Swindle* documentary?"

"Absolutely. Drivel."

Truscott nodded. "Right. You see, the media have left the public pretty confused on the debate. I've looked up some of the scientific papers, and they take a lot of effort to understand. The press reports a distilled version, over-emphasising some things, ignoring others, and most people think they know the terms of the debate. In fact they don't: what they are seeing is whoever manages to get the media's attention – the squeaky wheel gets the grease."

"Go on." Wilkinson was intrigued. This was exactly the way he saw it.

"See, the press is being lazy. They are allowing some-one else to define the debate. I want to understand what the real debate is, and present it without some preconception that someone is lying. What Durkin claimed he was doing, but actually going after the science, not a personal agenda."

"I see. Do you have a working title?"

Truscott shook his head. "No. The main point will come out from the science … if it's clear by the time I finish my pilot, I'll think of a title then, maybe … but be open to changing it if the facts don't fit."

"Very good. So what do you propose to do here?" Wilkinson was impressed – if this was for real.

"If you don't mind, I would like to go over some of the science with you, especially your work, then come back to-morrow to do an interview. You did say you could spare some time both days."

"Right. Let's start with some visualization." Wilkinson hadn't shifted from his seat, almost buried in the piles of papers. He now stood, surprising Truscott at his immense height. "Don't worry, I'm the tame kind of giant."

Truscott sniggered nervously. "Sorry. You don't look like your email."

Wilkinson smiled. "Let's go over to my grad students. They have better graphics equipment than I do. Probably to play games, but I never catch them at it. They claim they need the speed for their 3D stuff. Lots of cores, high-end graphics cards, stuffed to the gills with RAM ..."

They walked down the corridor, to a room with half a dozen students at PCs and Macs with various graphical models on display.

"Let's look at what Simon Wu is working on. Simon, this is Martin Truscott – he wants to do a doco on climate change. Can you show him your 3D visualization?"

"Yes, but it's a bit slow." Wu pointed at his machine. "What we are trying to do here is to see simultaneously the effects of a range of scenarios." He picked up a paper showing several IPCC scenarios, with the world coloured differently according to projected temperature changes. "With these scenarios, it's hard to see how the temperature changes compare. You can crudely see that this one –" he pointed –"is warmer around the poles than that one." He pointed again. "But it's hard to make an exact comparison.

"What I am doing is looking at ways of combining the data in three dimensions, so you can see the differences more clearly. As you switch scenarios, the temperature differences between scenarios appear as height differences." He fiddled with the program a bit, and bumps appeared sporadically, until the picture looked like mountainous terrain. "You can set the sensitivity, so the heights appear greater for smaller differences."

Eventually, it was possible to see what the program was illustrating, but it was painfully slow to redraw whenever Wu made a change.

"How have you coded this? Any ideas why it's so slow?" Truscott asked. Wilkinson watched with interest.

Wu explained. "Well, we have an 8-core Mac, and I have partitioned the work pretty evenly, and the renderer is using OpenGL. There must be some performance bottlenecks I haven't found yet, because it is hardly speeding up compared with running on a slower machine."

"Graphics card?" Truscott enquired.

"Best we could get for this model."

"You know, I've done a bit of graphics coding in a previous life. If you could let me have a look at your code overnight, maybe I could spot something."

Wilkinson shrugged. Intriguing ... not your average journalist. "Why not? Though I don't suppose it matters for your movie how long it takes to display. We can give you some screen shots showing relevant cases."

"Yes, that would be good," Truscott agreed. "I would like to see the upper and lower IPCC ranges compared, with a few scenarios like business as usual and a few of the proposals for CO_2 cuts."

Wu pulled up a copy of a recent IPCC report, and they discussed some options.

Wilkinson looked at his watch. "It's about lunch time. How about we leave them to this, and get some lunch? On campus is OK, but there is a pretty good selection out in the street. Do you have any preference?"

Truscott thought for a moment, then: "It's been a long time since I had a good laksa."

"I know just the place." Wilkinson gestured towards the doorway.

Before they left, Truscott pulled out a memory stick, and Wu copied his software over. "It's all in C++."

"No problem with that." Truscott reached for the memory stick.

"Some data files too – I hope the interface is obvious," Wu added, as he handed the stick back.

As they walked, each carrying a note pad, Truscott said, "Before we get to your work, I'd like your opinion on some of the 'sceptic' stuff."

"Such as?"

"Schoor at Harvard," Truscott offered.

"The sun man."

"That's right." Truscott nodded.

Wilkinson smiled tightly. "Let me throw that back at you: what do *you* think of him?"

"I had a look at some of his work. It looks reasonably convincing superficially, but he relies heavily on data smoothing techniques that, if I recall my stats right, are inaccurate towards the endpoints. If the IPCC models are right, the CO_2 effects should only have started to kick in recently, so his analysis is going to underplay those."

"Go on." Wilkinson was playing the Socratic tutor.

Truscott played along. "Then there's the question of the relative changes he models in solar irradiance versus temperature change. He doesn't anywhere demonstrate any physics

that would justify a claim that these changes would be significant. Right, that's my homework. Can I ask the teacher a
question now?"

"Touché. Go ahead."

They were outside the campus now, walking along a busy
street. Truscott fired away, dodging pedestrians. "How much
of a role, exactly, does the sun play in the IPCC's models? Is
it correct as some are claiming that the sun has simply been
left out?"

"No, that's nonsense." Wilkinson was vehement. "The
IPCC models include every natural and anthropogenic forcing
we can identify. The NASA people have some very comprehensive models – they really have thought of everything, including some details the IPCC in its collective wisdom thinks
are too uncertain."

"Forcing?" Truscott asked.

"Things that drive change directly. Secondary effects are
called 'feedbacks' . . . anyway, the solar effects, in all the calculations I've seen, can at most account for 20% of twentieth
century temperature change, and the most recent data shows
that solar irradiance is actually on the decrease."

They reached the restaurant, a bit of a dingy hole, but
Truscott didn't hesitate to follow his host, trusting in his local
knowledge. They went inside, and were directed by a hyper-
active waiter to the upper floor. There, their orders were taken
with dispatch, with no doubt left about how such a small place
achieved a decent turnover.

During the short wait for the food, Truscott pressed on.
"Some are claiming that the solar effects are delayed – so
a downturn over the last few years will only show up as a

downturn in global temperatures in a few years."

"Rubbish. Of course there are some delays in the overall system. If you heat the surface of the sea, it will take many years before there is an overall heating effect throughout. The oceans are enormous, and water is a poor conductor of heat. But temperature changes in the atmosphere are very quick, and surface temperature change of the sea is also fairly quick. How many years after the start of summer is the sea distinctly warmer to swim in?"

"Good point ... do you have physical models to back all this up?"

"Of course. I can give you some papers back at the office."

Truscott made some notes, and conversation died with the rapid return of the waiter with their food.

Truscott had forgotten how hard it was to eat laksa without depositing yellow spots on yourself. Somehow, the soup was especially attracted to white, and he had a mostly white shirt on. Managing the noodles was especially challenging – even one reasonably adept with chop sticks, as he was, could slop up a drop or two easily. Wilkinson watched through lidded eyes, while working on his own less taxing lunch.

Eventually Truscott had made enough progress to break for a question. "You said NASA has some details in their models that the IPCC left out?"

"Yes. The most important one is land-based – more specifically, Antarctic – ice melts.

"You may have noticed some are claiming that the IPCC has revised down sea level rise ranges since 2001. This isn't true. One of the things they've changed is that they separated out ice melts that they felt could not be modeled accurately. This in the view of many is being over-cautious. They should at least include the most conservative estimate that matches recent measurement."

"What has changed since 2001 that increased this uncertainty?" Truscott sat on the edge of his chair, the other half of his laksa forgotten.

"The previous models assumed that for each increment in temperature, you would see a similar increment in ice melt. In fact another part of the model put most of it back, through increased precipitation as temperatures rose."

"Hmmm. I read somewhere that ice caps were in fact increasing in mass ... "

"That's what happens when you stop reading before you reach the conclusion, because you've decided in advance what the conclusion will be. Read the paper again."

"This wasn't an original source, it was one of the 'sceptic' camp who reported this."

"Say no more. I can dig up some stuff for you, but the NASA people at Goddard are the real experts in this area and do a lot of other good climate work."

"Goddard?"

"Goddard Institute for Space Studies – you have to read their stuff to get a good understanding of what's going on. In New York ... "

"That's great. I am planning a US trip, and some of the people I want to talk to are in Boston – New York would be an inexpensive side trip."

"I hope you weren't only planning on visiting the denial bunch in Boston. Though I suspect you'll find a lot of the real scientists aren't too keen on talking to yet another journalist ... they've been misrepresented so often in the past. And since the Nobel, a lot of them have been taking the position that the time for debate is long past."

"Well, several of the people our local sceptics like to quote are in Boston, and this is only a pilot I'm making, so I am not looking for universal coverage. I was really hoping to get at least one of the mainstream IPCC bunch at MIT or one of the other Boston places, but NASA sounds like a good option. I'll add it to my list when I start trying to find contacts. Anyone I should look for there?"

"The best ice man there is Geoff Blunt – if you can talk to him, you will get the real scoop on this. I have some contacts who may know people there – maybe I can help you set something up."

"Thanks, I'd really appreciate that," said Truscott.

Wilkinson made a note on his pad then looked up. "But do yourself a favour: go to `realclimate.org`. A lot of this debate is already covered there, and you can find plenty of pointers to papers and hence hints on whom to talk to there."

"OK, I'll check that out this afternoon."

"Oh, and do yourself another favour."

"What?"

"Finish your damn laksa. It's getting cold."

As they walked back to the campus, Truscott looked thoughtful. "Look, I think I have enough to start working up some questions, but I want to explore `realclimate` as well. Do you have a space where I could sit for a while, with Internet access? Otherwise I will have to go back to my room where I can work, but I don't want to lose the train of thought now I've started."

"I know what you mean. I think the quickest will be to let you log on as one of the students – there's always one or two not in the office and one of the others can fix you up – as long as you don't tell anyone. This place is full of bureaucrats." He winked; noting Truscott's lack of reaction, he guessed his visitor had never had the pleasure of seeing the business end of a university. They entered the campus, leaving the city street behind them.

*   *   *   *   *

For the rest of the afternoon, Martin downloaded papers, read articles and arguments on `realclimate`, and made copious notes.

As he headed back towards his hostel, thinking about all the papers on his memory stick, he remembered another piece of unfinished business. "How long has it been since I did some serious coding? Let's see if I've still got it."

He stopped at a fruit store on the way, and bought an apple and a banana. "That should keep me going for a while."

In his room, he opened up his computer, and transferred the software over. He opened up the student's code and peered at it for a while. "No wonder this doesn't perform. Let's see what I can do here. Group this bunch into an object. Pad out the memory to span a whole cache block. Change the order here so all related data is close in memory. What's this lock for? Don't they know locks don't scale?"

He started shifting code around, changing data structures, tweaking details. He felt like a fish back in water. Close to midnight, he was ready for his first test run. Even on his much slower Mac, the code flew compared with the original. He tried a few more tests, then, satisfied with a day's work, fell down on the bed.

"Man, it's so good to know I can still do that. Let's see how it runs on some real iron tomorrow –" he looked at his watch – "later today."

He went to the bathroom, cleaned his teeth, took a quick shower, and collapsed into bed, falling asleep faster than any time he could remember.

* * * * *

Morning – and Martin's demons came crowding back with a vengeance. The familiar feeling that Julie should be there followed by disappointment, the cruel memories of how she had gone out of his life like a flick of a light switch. Then his dad: why had the old man told him a pack of lies about mum? Wasn't a boy entitled to know about his mother? The anger after all this time – when had he last spoken to dad? – was dulled. But he was entitled to *know*.

He lay in bed contemplating for a few minutes, then tried to shake off the mood. Coding. It brought back so much. The unhappy days, hiding in computer games, imagining himself as the world's greatest game hacker, the feeling of achievement when he did something really cool, the lows when he didn't know what it was all *for* . . .

He shook off the mood. "There's work to be done." He got up, pulled on some clothes and started reading papers. After half an hour or so, he grinned. "My therapy. It works. What does Prentice know?"

Despite reading through a good selection of papers, writing out a few notes on lines of questioning and a leisurely breakfast from supermarket supplies, he was early for his appointment – 9 am for 11 am. Late night and short sleep or no, his desire to see his code perform had him wide awake. He found the student office. Wu wasn't there yet, but one of the others was there. "Sorry, no one introduced us yesterday. This other guy, Wu, was showing off his software, and I made a few tweaks to it."

"I'm Mike. I can't log on to his machine but we could run it on mine. I'd like to see what you did."

Mike took the memory stick and copied the software over.

Martin leaned over and pointed, as Mike hunted over the icons. "That's the new version."

Mike launched it and opened a couple of data sets to compare. "Wow." It rendered so fast, there was no perceptible delay.

Martin grinned. "I did see some speedup on my machine, but I only have a dual core, so it's nice to see it really works on a decent machine."

At that moment, Wu walked in. Mike called out to him – "Over here." Wu was speechless as Mike demoed Martin's version. Wu then took charge, and tried a few variations.

"I can't believe it. I spent months on this, and you've made it so much faster in just a few hours. Let's see if Dr Gene is in."

They went to Wilkinson's office.

Wu knocked on the door, which was ajar. Wilkinson appeared to be finishing his breakfast, and wiped crumbs off his face. "What is it?" He then saw Martin. "Haven't you heard of being fashionably late? You're nearly two hours early."

Wu interrupted as Martin was about to explain. "You've got to see what he did to my program."

Wilkinson's eyebrows rose, in response to Wu's excitement. "OK, let's see what he did."

They trouped over to the student office, where Mike was still playing, with a bigger group of students surrounding the screen. Wilkinson shooed them away so he could see.

"Amazing. And is this all totally correct?"

"I didn't have the information to check the physical model fully," Martin admitted. "But I ran a bunch of tests against the original, and it all looks correct. I think your students are more qualified than me to check the physical model anyway. I documented all the changes, so it shouldn't be too hard."

"More qualified? Where did you learn to code like that? They've been working on this for months, and I couldn't find any major flaws in their approach."

"It used to be sort of a hobby of mine ... but I did do a bunch of courses in computer architecture, operating systems and algorithms. I tried to pay attention to how everything fits together, not the usual forget stuff after each exam."

"Hear that guys?" Wilkinson turned to his students. "Listen and learn."

He turned back to Martin. "We are pretty early for our appointment, and I have another in fifteen minutes. Why don't we chat a bit in my office?"

Martin nodded, and Wilkinson led him away.

He ushered Martin to a seat, and closed the door behind him. "Forget the interview thing for now ... What you did there wasn't just hobbyist tweaking. My students are good.

Maybe they would have fixed the performance glitches in a week or two, but you did it in a few hours.

"Don't take this as a bribe or anything to do a better interview, but if you ever want a change of career ... have you thought of doing a PhD? We are doing some really big models, and the performance challenges will be much greater than this toy example."

Martin shook his head. "For a start, I didn't do an Honours degree –" in Australia, many undergrad degrees take only three years, and the Honours year is an extra degree, usually the entry requirement for a PhD – "and even if I did, my GPA wasn't that great in my later years."

Wilkinson put his hands up in a stopping gesture. "These are all filters but they are not hard rules. We very seldom do this, but we can accept a PhD student without a four-year degree in an exceptional case. You can handle the computational side better than most students with a climate science background, and we can make a case to the bureaucrats that your documentary work is more than equivalent in intellectual scope to another year of study.

"Think about it anyway. I'll check with a few people to see if we can swing it, because I've never handled a case like this before, but I will push your case if you want to do a PhD with me."

Martin pursed his lips, then nodded slowly. "As you say, we can't make this a bribe for a softer interview ... but it's tempting. It's not like I have a regular income. But I haven't studied for five years ... "

"What do you mean, haven't studied? What are you doing to put this thing together? You're reading papers, arguing with the experts, putting together a coherent case. Believe me, most academics would kill to get a student like that."

"Yeah. Well, let's see if you still like me after the interview." Martin smiled.

"I admire your spirit." Wilkinson glanced at his watch. "Not much time before my next appointment. See you at eleven."

*   *   *   *   *

Martin was waiting outside Wilkinson's office at eleven, and Wilkinson rushed back no doubt from some other appointment, a few minutes late.

"Sorry about that. Do you want to do the interview here or outside?"

"Outside, if that's OK with you – I prefer natural light."

They found a quiet corner near the building.

Martin began with a general question. "The sceptics claim it's all just bad stats – that if you fudge the data in a statistical model, you can make it fit the conclusions."

"Well, that would be a good criticism if that were true. Unfortunately this is hard to illustrate in an interview, but you've looked at my students' code – I hope you noticed that it's based on modeling the physics, not on statistics."

"That's correct. I did examine some code in detail and it definitely was mostly based on physics, not statistics ..." Martin pressed on. "What about the point some raise, that we can't predict tomorrow's weather with certainty, so how can we predict what the climate will be like in a century?"

"Different time scales, different kinds of prediction." Wilkinson settled into a well-rehearsed speech. "Weather is a

specific thing, here and now. Climate is a general thing, averaged. You don't know if you'll get the flu tomorrow, but
we know that over winter, a certain fraction of the population
will normally come down with it –" he illustrated fractions
with his hands – "and that every few decades, a really virulent strain arises. Should we stop working on precautions
against virulent strains of flu just because no one can tell you
if you will be sick tomorrow?"

Martin changed tack. "I see. Another point some have
raised is that temperatures peaked in 1998. I see now 2005
is considered to be in the same ball park. Why should 1998
have been warmer than subsequent years?"

"1998 had a strong El Niño – an event that is unrelated
to global warming. 2005 was at least as warm without an El
Niño."

"What causes an El Niño?"

"The exact cause hasn't yet been pinned down. One recent observation is that an El Niño typically follows three
years after a volcano in the tropics. But that's a statistical
study, and we really need to understand causation ... in any
case, it's a temporary phenomenon, a blip in the long-term
trend."

Martin nodded, and decided to move on. "I've seen some
claims that contrary to predictions of warming, Antarctic ice
is thickening."

"That illustrates the danger in stopping when you find
data that suits your case. The Antarctic ice caps are thickening in the centre because of increased precipitation, which
is exactly what the models predict. They are getting thinner
at the edges, at a rate faster than the models predict. That to

me is cause for concern."

Martin pressed on. "What about the claims some are making that all changes are a result of changes in solar activity?"

"Well, they have to justify the data they are coming up with, which doesn't match what the rest of us have. Our data is widely cross-checked. We have strong confidence in its validity. If you look at the IPCC reports, you will see that they separately model natural forcings – solar activity, volcanoes – then add human inputs. Neither the natural nor the human inputs are a strong fit to observed temperature changes. Combine the two, and you do get a good fit."

Martin brought in McCarthy's line: "What of the argument that CO_2 is getting a bad rap – its concentration is in parts per million, and water vapour is a much bigger contributor to the greenhouse effect?"

"Sound logic, poor conclusion. The thing we are worrying about is not the total greenhouse effect, but the increase. Without any greenhouse gases in the atmosphere, the average surface temperature of the Earth would be over thirty degrees Celsius lower. Now, tell me who is driving temperatures up by creating vast quantities of water vapour? Has anyone measured a substantial increase in atmospheric water vapour? In any case, water vapour has a relatively short occupancy in the atmosphere. It precipitates out when temperatures drop. CO_2 increases on the other hand persist for a long time, possibly centuries – the carbon cycle was more or less in balance before industrial times and lowering temperature doesn't reduce the atmosphere's carrying capacity, as it does for water vapour."

"Doesn't CO_2 dissolve in water?"

Wilkinson nodded. "Quite right. However, the oceans' capacity for storing CO_2 is reduced if they warm up, so the net effect of any warming caused by anything else is more CO_2 vented from the seas. That in fact is what we believe happened in previous warmings. The Earth warmed because of changes in solar heating ... the long term effects of the planet's orbit ... then the sea vented CO_2 which increased warming."

"And that's happening today?"

"Yes, but as a consequence of warming caused by anthropogenic CO_2 ... "

"How can we be sure of that?" Martin pressed the point.

"First of all, as I explained before, natural forcings don't explain the changes we're seeing. Secondly, we can determine by isotope analysis the source of CO_2 in the air – whether it's from coal, for example."

"What fraction then of climate change is it reasonable to put down to natural effects?" Martin asked.

"Possibly twenty percent – it's unlikely to be significantly more than that. Earlier in the twentieth century, the sun was the dominant factor, but not any more. Recent studies have in fact shown a reduction in solar effects."

"How certain then can we be of the effect of an increase in CO_2?"

Wilkinson launched into lecture mode. "The physics says if you double the CO_2 in the atmosphere, you increase energy absorption by four watts per square metre – for the less technical, this means we have a specific calculation which says how much the temperature rises as we double the CO_2 in the atmosphere – the models show a range of 1.5 to 6.2 Celsius."

"Why the range of values?" asked Martin.

"If we are looking purely at forcing in an idealized situation, we can come up with an exact number. This range of numbers represents *climate sensitivity* which has to take into account feedbacks ..."

"Could you explain what you mean by 'forcing' ... and 'feedbacks'?" Martin asked – worrying that the technical detail could lose the audience.

"Yes. A forcing is a trigger event – something which on its own causes a change. We distinguish that from an indirect effect, which may either increase the change – a positive feedback – or damp the change – a negative feedback. Many people raise the issue of water vapour as a greenhouse gas. Water vapour is never a forcing because it doesn't stay long enough in the atmosphere to trigger climate change. But if temperatures increase for another reason, the atmosphere's capacity for holding water vapour increases."

Martin nodded. "Any other examples of feedbacks ... ?"

Wilkinson waved his arms expansively. "There are many indirect consequences. For example, as you increase temperatures, you lose ice. Ice is much more reflective of heat than water, so this leads to a positive feedback. Less ice means more warming, which adds further to ice reduction. And more warming means more CO_2 released from the sea ..."

"Could the feedbacks from increasing CO_2 lead to a runaway effect ... turn the Earth into something like Venus?"

"Good question. Venus's lower atmosphere is over 90% CO_2, and the surface temperature is hot enough to melt lead. We have calculations that show such a runaway effect is extremely improbable under any scenario you could conjure up for Earth.

"But we could certainly get as far as making life very difficult for humanity – bigger sea level rises than in the IPCC models, extinction of important food sources ..."

"And how likely are those extreme scenarios?"

"We can't be sure – a lot depends on cutting back CO_2 rather than rapid growth of CO_2. The real experts in the supposedly more extreme scenarios are at NASA."

"You mean the ice people at the Goddard Institute for Space Studies?" Martin picked this up from the discussion of the day before.

"Yup. The IPCC has not yet included rapid ice melts in their scenarios, because they want a detailed model for everything in their reports. Just because we don't have a clear understanding doesn't mean it can't happen."

"Thank you, Doctor Wilkinson."

Martin turned off the camera. "I'll have to see how that goes with interviews from NASA and the contrary crowd. I think you mostly put it pretty clearly for a non-specialist audience. Thanks very much."

"And the PhD?"

"Let me finish this first. It has some appeal – but it would be a big change for me." Martin smiled tightly, covering his emotions. Would a PhD be even better therapy? A move to Sydney would certainly feel like a new start to life ... and having his skills valued like this was something to get used to.

4 MIT

BOSTON'S LOGAN AIRPORT is pretty busy considering that Boston is a modest-sized city. For a stranger, navigation can be confusing, especially if on a tight budget. The *Lonely Planet* advises taking the *T* – the underground system – rather than a cab, because Boston cabs were notoriously over-priced. And, in any case, everything is pretty reachable even on foot once you are in the city centre, which is pretty close to the airport. Martin took this advice, then struggled to make sense of the terse train announcements. Jet-lagged, tired and balancing a heavy backpack, he battled to read the train route maps. Eventually, he found his hostel – the Blue Parrot – then, to his horror, found he had barely thirty minutes to make his first interview – Jim Hayes at MIT.

Damn Hayes's busy schedule ... why couldn't the stupid airline sell seats at a decent price on the dates when I want to travel ...?

Martin rushed back out hoping to catch another *T*, working up a bit of sweat despite the mild temperature. Luckily there was one waiting as he hit the platform and even more luckily, it was going the right way. Minutes later, he emerged from the red line stop at Kendall Square, and tried to orient himself. He pulled a campus map out of his pocket, and, with a few false starts, found himself in the right place.

MIT sends out mixed messages – all steel and glass, yet somehow with an air of older money than it deserves. Hayes didn't fit the image of the MIT that Martin knew from his computer science studies – new tech, challenging the boundaries of cool. Instead, he looked pretty conservative, with what looked like an old school tie and neatly pressed slacks. His office was neater than average for an academic's with tidily shelved books and boxes of papers.

He had a modest-sized screen on his desk, the computing parts hidden somewhere out of sight. As Martin approached the office, Hayes spotted him, and jumped up to greet him. "You must be Truscott. I cleared a bit of time so we could get all this straight. That Durkin fella really spoilt the case by doing such a sloppy job."

"Good, that's what I'm here for. Would you like an informal chat before we record? I am doing a pilot, but if any of it is good enough, it could still go into the final edit. Either way, I'll let you review the final edit for accuracy and context."

"Excellent. A short chat would be good, then we can go to a brighter spot, since I see you haven't brought any lights with you." He gestured towards a chair.

"Thanks," said Martin, sitting down and parking his gear on the floor, except his computer, which he kept with him. "To start with, could you outline your problems with the IPCC approach?"

"Yes." Hayes moved to the edge of his chair. "My main problem is that they are presenting the science as settled, when there are significant things we don't know."

"Do you mean that what is going into the model is mostly sound, but they haven't included some details that should be there? Or that the approach is not sound?"

Hayes spread his hands. "More the former. There are areas where we don't have good models."

"Such as clouds," Martin suggested.

"Exactly." Hayes nodded eagerly.

"Could you elaborate?"

"Water vapour is a greenhouse gas. On the other hand, clouds reflect heat. The combined effect is very hard to model."

"Have you been able to quantify the errors in the IPCC models?"

Hayes grinned. "This is sounding more like a scientific discussion than something for TV."

"That's the purpose of the preliminary discussion. Should we go to recording now that we have the approximate terrain, and I'll try to word it more for the public?"

"Good with me." Hayes stood up, looked for his coat, then decided better of it, remembering the weather was warming up. "Let's go outside."

They found a convenient spot, with the Charles River and Memorial Drive as backdrops.

"That looks good," said Martin. "Let's do it here."

He set up the camera, then moved closer to Hayes.

"Let's get your problem with the IPCC straight for a start."

They more or less repeated the conversation up to the part about the cloud problem. Martin continued: "OK, so the

IPCC hasn't adequately handled clouds in their models. Can you point to anything in their predictions which has come out wrong as a consequence?"

"It's not that simple. The IPCC model takes into account many factors, which result in a range of values – across all scenarios, temperature ranges were predicted at the turn of the century to be anywhere from 1.4 to 5.8°C– a bit less in the latest models. You can't just look at one factor in isolation. But if you haven't modeled that factor accurately as part of the whole, the errors could be bigger than you're predicting."

Martin pressed the point. "So you're saying the IPCC modeling is essentially correct, except some details are missing?"

"That's overstating the case." Hayes shook his head for emphasis. "The IPCC models are not totally incorrect, but we don't know the extent to which the missing detail makes the predictions inaccurate."

"What about the effects then of anthropogenic warming, the effects of CO_2? Aren't those significant anyway?"

"Possibly. But the failure to model clouds adequately means we could be vastly overstating the effects due to CO_2. More CO_2 could cause some warming. This could cause an increase in clouds with unpredicted effects. I have some models which suggest that the net effect could be cooling."

"Do you actually have numbers showing that the CO_2 effects could be overstated?"

"No. I don't think anyone can come up with such numbers, because a major element of the system is missing. My belief though, based on my own research, is that it is extremely doubtful that there will be significant anthropogenic warming by the end of the century. According to my long-range studies, climate variability is much less than the alarmists would have us believe. It's a self-regulating system. Temperatures increase: you get more water vapour. More water vapour means more clouds, which reflect heat. Go either way, and you reach equilibrium quickly."

Martin nodded. "I see in some of your earlier work, you said that your models predicted that during an ice age, there wouldn't be global cooling – that the tropics would stay warm. I don't see that mentioned in your more recent work."

"Some more recent measures contradict this finding: there is a claim now out there that the tropics were much cooler than in my model. Unfortunately this is not my data, and I don't have the resources to check it, but I am sure it's wrong."

"A sceptic might say that's rather convenient – your theory no longer fits the data so ... "

"Well, that would be unfair," Hayes said vigorously. "That work was from my distant past, and I've moved on to other things. If I were still actively working on those models, I would look in more detail at correcting the data."

"OK." Martin decided to move on. "Another thing has been puzzling me – you've mentioned the case of scientists hyping up global cooling in the 1970s, but I haven't been able to find a large body of scientific papers making that sort of claim – certainly not with the degree of confidence we see in climate science today. One paper I found for instance predicted glaciation some time in maybe 20,000 years – yet it is widely cited by others making the claim of a cooling hysteria."

"That one was stopped in its tracks when temperatures were clearly rising by the late 1970s." Hayes looked a bit annoyed.

"Maybe so, but is it really accurate to call something 'hysteria' when a handful of papers covered the matter, and then mostly with serious caveats?"

"Well, all I can say is that I was there at the time, and I've seen both debates."

"Thank you professor." Martin turned off the camera. "I'll have to think about that a bit … try to make sense of it. I'm going to Harvard tomorrow –"

"– to see Schoor?"

"Right. And there are a couple of other people around here I should look up to get the other side."

"You'll get a lot of that here." Hayes smiled wryly.

"Thanks for your time, professor."

"That's Jim – no need to be so formal, though I suppose professor sounds better on camera. More authoritative. Though it doesn't stop me disagreeing with a whole lot of other people who call themselves 'professor'."

Martin nodded. "I appreciate your time and your helpfulness."

The interview over, Hayes headed back to his office, while Martin lingered over the view, before gathering up his equipment and heading for a coffee shop.

A quick double espresso later, he was – despite the caffeine shot – about ready to crash, so he gathered his stuff, and made for the Kendall Square *T* stop, and found his way back to the hostel in a bit of a daze. The rest of the day was a confused mess of replaying his mission and memories of better days, when he'd been an undergrad, with high hopes of setting the computing world alight.

Better days? Who said nostalgia was the result of bad memory? There was an empty space in his life called Julie – that had seemed a happy time but was it? If she could have walked out on him like that, was any of it real? He drifted off, and woke with the same thought.

He went through his morning routine, then tried to make a few contacts with the mainstream camp. All were too busy. Probably an excuse ... maybe Wilkinson was right: they saw the debate as closed. He spent much of the rest of the day walking off his jet lag ... what did I discover in that melatonin exposé? Oh yeah ... a day in the sun is pretty good for getting in the time zone ... so a recovery day isn't the worst thing ....

Oh well. For the pilot, it will have to be Wilkinson and NASA giving the other side ... I just hope I can get more for the full thing ... if there's a full thing ...

That night, he had a vague impression of the other hostel inmates, as he confronted a keep-awake snack of dried fruits and nuts at the kitchen table.

5 Harvard

CAMBRIDGE, MASSACHUSETTS . . . for the US, it has an old-money feel even if it's quite a few centuries short of its British namesake. Harvard, as one of the top universities in the world, somehow has to be catching up, positioned as it is in a location named for one of the oldest universities in the world. MIT escapes that feel because of its new age high-tech image, but Harvard . . . it tries so hard.

Martin woke early in the unfamiliar time zone, and took a stroll through the streets, taking in the scene of coffee shops opening for business, early risers catching the first coffee and donut of the day, the croissants and other delicacies appearing at the counters. It was a fine late summer day, a couple of weeks to fall if you were American, with red leaves starting to appear on the few trees in the vicinity of Harvard Square.

He'd missed all this in the MIT visit, the crazy rush from the airport, finding his hostel, finding MIT. This time, he was going to pick up some atmosphere. Another day, another pace.

He settled on a small non-chain coffee shop, and ordered a cappuccino and pain au chocolat, wondering how accurate his pronunciation was – but if the locals had as much difficulty with Australian as he had with Bostonian, pointing probably did it more than whatever he said. Whatever. The chocolate croissant was buttery and flakey, with a hit of molten chocolate, and it went down really well with the ultra-hot cappuccino. American mass-market coffee may be dreadful but the small coffee shops in Boston and New York really knew their art, if the web sites he'd reviewed were to be believed ... so far, so good.

He checked his watch. Two hours to the meeting, plenty of time to consider finer points, and review the literature. He opened his backpack and pulled out a few papers.

"Let's see now. This one correlating solar irradiance with temperature change looks like the key one. Durkin used this somewhere. I should start with that."

He sat there for a moment contemplating, and saw someone else using a computer. "Maybe a wireless hot spot?" He opened his backpack again, and hauled out his own computer, and flipped it open. Sure enough, the WiFi icon was showing activity, and an offer to join a free network popped up.

"Might as well look up some more detail."

He went to `realclimate.org` and started to search for Schoor and Harvard, in various keyword combinations. There it was suddenly: a new angle. A member of Schoor's group, Sherry Lee, had been very active in debunking the ozone hole. Then, when the evidence was overwhelming, switched to debunking climate change. And one of her subsequent papers claimed that climate change was caused by the ozone hole.

Martin laughed out loud, some of the generally taciturn Bostonians looking up from their morning papers briefly.

"Unreal! Why hasn't someone picked this up before? Could they be a bunch of shills, Harvard or no? Let's see who funds them."

He switched his focus to searching for things on the Harvard web site. Sure enough, the papers `realclimate.org` had referred to were all there. He then took a closer look at the paper McCarthy had told him to read. Now, this was the McCarthy who said one can't model complex physical systems statistically, you have to look at the observational data and the physics. The paper was essentially about finding formulae to join dots, and make one lot of dots join more convincingly than the other. Even though he had read the paper a couple of times, on remembering McCarthy's words, a thought struck him. "So, fine. If you look at the biggest change in the solar graph as a percentage variation, then compare it with the percentage variation in the temperature, is it close?" He did some mental arithmetic, then verified it on the calculator program on his computer. "Nope ...nowhere near close. What did Wilkinson say? Two percent increase in solar irradiance per degree temperature rise ...That's another question, then. I think this time, I'll go straight to camera."

Time was marching, but he still had an hour. Martin closed his computer and repacked everything. "A walk will clear my head." Mass Ave (Bostonians, it seemed, abbreviate everything) was starting to get busy. He took a stroll over towards the Charles, and watched activity on the river for a while. He had to cross the river for his appointment; contemplating Harvard from the other side of the river helped somehow to put things into perspective.

"This is just one interview of many. Harvard is a big

name, but even the best universities have weak links. Academic freedom. Once you have tenure, you can do anything. Or at least, that's the way it used to be ... maybe still is if you have a name like Harvard."

Finally, he picked everything up again, and walked across the bridge. He still had a few minutes, but he had to find the right building. The old navigation challenge ...

With a minute to spare, he was at the right door. It was closed, unusually for the university offices he'd visited so far. He paused a few seconds, then knocked. A muffled voice issued through the heavy wood. A few seconds later, the door cracked open, revealing a stuffy-looking office – despite the high ceiling, it somehow felt cramped.

"Martin Truscott. Here to do an interview on climate change or the contrary." He smiled, trying to break the ice.

The occupant of the room pushed the door open further.

"Schoor." Obviously he was a man of few words. He stood aside to let Martin in. Schoor was short and tending to fat, with thinning hair. He had an unusually sartorial suit for an academic – somehow he contrived to look shabby despite wearing expensive clothes.

"Look, the lighting outside is great, and it would make for a suitable backdrop if we could see some Harvard buildings behind you while we talk."

Schoor grunted, and sidled out of the office, locking the door behind him. "Wonder if the whole interview will go like this," Martin subvocalized.

"Hmph?" queried Schoor.

"Nothing, nothing. Let's find a nice spot outside."

He steered Schoor to a position where some red leaves

were visible, adding colour to the otherwise drab scene. He set his camera on a tripod, and walked over to Schoor. He perched his computer on a handy bench, ready for pulling up any reference material.

"Now, do I need to go over why I am here?"

"No. No. I remember the emails. You want to do another *Swindle* show. McCarthy sent you to me. Let's get on with it. I have work to do."

"Right ..." A pause to allow editing out the preamble ... He introduced Schoor, then: "A big part of your recent work has been correlation of solar activity to temperature change, is that right?"

"Correct."

"Now, let's see if we can do this for a non-technical audience. McCarthy told me that a big problem with the IPCC's science is that you have to base this all on observation and physics, not just statistical models. Do you agree?"

"Of course."

"Now, your solar work ... I've read a few papers, but it seems to me that it is all statistics. Curve fits and the like."

"Yes, but based on observational data."

"Right, but the way you smooth the data ... If you do a ten-year moving average for example, that means your analysis pretty much ignores any recent changes, doesn't it?"

Schoor was getting annoyed. "I'm a Harvard professor. I don't need a journalist to lecture me on data analysis techniques. Do some background reading and come back to me."

"We'll leave that aside then. Another thing that interests me is that you are claiming that solar variations are a much better fit than CO_2 variations to temperature change."

"That's what the data shows."

"Yes, if we accept the analysis technique – but what I am more interested in here is whether you've worked through the physics. If the solar irradiance changes by the percentages in your papers –" he held up one for Schoor to see – "is that change sufficient to account for all the temperature change?"

"Well, do you think it is?" Schoor snapped.

Martin tried to calm things down. "I'm sorry, but it's your work. I am just trying to determine ..."

"Become a referee for a major journal, and you can ask that sort of question next time I have something to publish. This paper has been published so my peers – of whom you are not one – were satisfied. Do you have anything else? Time is marching."

"Sherry Lee is also in your group, isn't she?"

"Yes, what of it?"

"She recently published this paper –" Martin brandished it in the form of his computer – "claiming that the ozone hole was a contributory factor to global warming."

"Nice piece of work. One of the most productive members of my group."

Martin didn't miss a beat. "Well, I looked at some of her earlier work, and she was pretty active in publishing papers claiming that the ozone hole wasn't real."

Schoor started to collect himself. "That's it. I wasn't expecting a grilling."

"Oh and one more thing. Is any of your funding *not* from fossil fuel sources?"

Schoor stormed off quivering with rage, just in time for Martin to reach the camera and turn it to follow him.

6 Blue Parrot

THE BLUE PARROT had the look of a grand old house gone to seed, every room converted to sleeping space, except a few held back for rudimentary bathrooms. It was built in the old style, with face brick on the outside, and thick, solid walls. Like most hostels, it was moderately clean but not spotless.

Martin had a room with four beds, two of the others occupied, details coming into focus now that he was less preoccupied with staying awake and making it to appointments.

Therapy? He hardly had time to think of that. So it must be working.

When he got back from Harvard, he found one of the other occupants of the room there, snoring loudly. His jet lag still had him in a wakeful state, despite feeling tired, so he went down to the TV room with his computer and camera. There was no one there, so he decided to offload the day's work for review.

The room was shabbily comfortable, with overstuffed chairs, and a giant TV. In a corner, there was an ancient computer for email for those who didn't bring their own computer with wireless. Martin had stopped marvelling at the fact that internet access was so cheap in the US even compared with Australia's relatively modest costs, that it could be offered free even in a $29 a night hostel.

He sat on one side of a couch, to make some work space, connected a FireWire cable, turned on the camera and opened the computer. He searched in his backpack for the charger and an adapter to fit the US power point. Everything set up, he transferred the movie over, and – containing his keenness to review the interview – backed it up to a DVD. He waited impatiently while the burn completed. Finally happy that he could edit without risk of losing the original, he opened up iMovie. Not for the first time, he was thankful that an entry-level Mac allowed a vague approximation to professional editing at no extra cost.

He pulled in the movie, trimmed off some of the start, and played it through. He sat back at the point where Schoor had gone storming off. What to add . . . a clip showing the acknowledgments at the end of a Schoor paper, showing only fossil fuel funding? Did it need more than that? No. Overkill would spoil the message. Sometimes no comment is better than driving the point home with a sledgehammer.

He played the clip again.

So engrossed was he in his handiwork, that he didn't notice someone else was now in the room – until a delicate touch on his shoulder almost had him jump out of the couch – a hard task, considering how deep he had sunk into it.

"I'm so sorry – but that looked so interesting. Was that you doing the interview?"

"That's OK – I shouldn't be editing in a public place. I'm

working on a TV show about climate change – could the sceptics be right, or are they talking rubbish?"

"Rubbish, I'd say – if that's the best they can do."

"Yeah, well, that was Schoor at Harvard. He didn't take
my questioning too well. Some of the others I've spoken to
haven't been so bad. But the case is looking a bit flimsy. I'm
going next week to New York to talk to some NASA people
to get the other side."

"But tell me, how do you get to be doing such a low-
budget TV show? I mean, staying in a dump like this, doing
your own editing . . . I'm doing this because I'm a student, and
this is what students do."

"It's a long story, not too exciting."

"Try me."

He looked directly at her the first time. She had an en-
quiring face, curly dark hair, an athletic look. A serious face,
but one which looked capable of fun. He could still feel her
touch on his shoulder, even though the hand was long gone.

"Sorry, I am not so hot on the social niceties. I'm Martin.
Visiting from Australia."

"So I guessed from the accent. Angela. From California.
Angie to my friends." She flashed a slightly naughty smile, a
hint of dimples. There it was, the capability of fun. He found
himself noticing details he hadn't been taking in with other
people.

He patted the couch. "This could be big enough for two
–" not sure if this was the right move.

She smiled again, and sat next to him. "Your story."

"Well, I was a pretty antisocial or I suppose asocial kid.
I didn't really have friends at school. I liked to be different,

and you know what kids can be like." She nodded. "I got into computer games at an early stage, blood and splatter." He grimmaced. "I can't imagine doing that now. But I was so into it, I really wanted to get into doing my own games. I researched what went into building a good game, and decided I needed a good hit of physics and calculus, layered on computer science. Most of the kids who are into games don't know half of what goes into them. They see the cool, and that's it.

"So there I was a couple of years into my degree, the hard physics and calculus grind behind me. I was getting into the more advanced computer science, and working on what I thought was the coolest game of all time in my spare time. The games companies all say they value experience over education, so I had to have something to show off. I didn't entirely believe them: you need physics to make a ball bounce realistically or blood to splatter ..." He stopped at her reaction. "Look, this is who I was. It's been a long journey from there."

"Sorry, I was just thinking you looked so nice, then you said that. Go on, I want to hear what changed to make you nice."

He grinned. "I don't know if there is something so obvious. I was going on like that, my whole life planned out and everything. I knew getting into games was not dead cert, but computer science jobs were easy. There was the dot com boom, the Y2K mania – everyone was either doing new stuff or fixing the old stuff. So the plan was either get into games straight off, or get into software and keep doing games in my spare time until I was noticed by one of the studios.

"Then I was noticed, but not the way I expected.

"I met her – Julie …" he said the name softly, as if he wasn't confident of pronouncing it right. He realized he had not said the name aloud for years. "It was some new agey tree hugging thing, I forget what. After hours on campus. I went because there was free food. You know what students are like." This time she nodded vigorously.

"You did say you are a student, didn't you? You related to that more than the 'you know what kids are like' line."

"Guilty as charged. I've just finished off undergrad at Berkeley, and looking into getting into MIT for a PhD. I wanted to scope the place out and try to pick some hints on the right buttons to hit."

He looked puzzled. "If you've finished at Berkeley, why are you only applying now? Can't you get accepted subject to final results?" He tried to remember the schedule from when he was looking at PhDs, and this didn't feel right. Australia, with its southern hemisphere calendar, didn't synch well with the US – that he remembered.

"There were some really great courses that I didn't want to miss, so I'm doing an extra semester. Crazy, I know. I'm going to be doing two years of courses at grad school anyway, but I'm going to miss my old gang – wanted to hang out with them another semester, those that didn't finish up and leave. And anyway, MIT has February entry, so I'm targeting that. What about you? Did you ever think of grad school? Isn't that what all the good CS grads do when the jobs are thin?"

"Well, I did think of going somewhere like Berkeley or MIT, but a straight-A student a year ahead of me applied to MIT and didn't make it. I wasn't a straight-A student. I had

... distractions. I asked around at home, and was told with my grades, I shouldn't bother to apply."

"If that interview is anything to go on, it's MIT's loss."

"Flattery." He grinned. "How did you guess my weakness? So what will you be doing in your extra semester?" he went on quickly.

"Not sure yet – maybe some extra CS, a few more bio courses. Maybe creative writing. But you were telling me *your* story."

"So ... distractions. I met Julie at this tree huggy thing."

*   * . *   *   *

It was a balmy summer day, enough humidity to enervate, but not so much as to put you in a total sweat. Brisbane was like that: just a little worse than comfortably hot. The talk had been boring, obvious stuff about over-farming or over-fishing, or maybe both. He was a bit hazy on the detail because he had almost dozed off.

Finally the talking was over, and the refreshments beckoned. Fine cheeses and a glass of wine were an adequate bribe for the boredom. For once, he wasn't totally focused on getting back to coding, or trying to make sense of where calculus fits into making a game play like reality. He took his time, picking out nice pieces of cheese, fruit and crackers. The wine wasn't half bad either. Not that he knew much about wine: it was a bit out of his price league, once he'd bought top-of-the line graphics cards and extra-speed memory.

He was chewing thoughtfully when he felt a light tap on

the shoulder. His eyes lost focus for a second, then focused more sharply than he thought possible.

"Hi. I'm Julie. I don't think I've seen you at one of these before."

"Mmph. I uh only recently became aware of the whole environment thing."

"Well, maybe I should educate you a bit. I've got some stuff in my room – I'm just a couple of blocks off campus – if you can spare the time."

His socially undeveloped mind suddenly took in that this was a pick up. "Yes, yes of course. I can't stop now that I have the basics."

They walked the few blocks which turned out to be a 15 minute walk. The wine had an unexpectedly strong effect on him, and he found everything a little blurred, making that sense of sharp focus when he first looked at her the more distinctive.

She had long, slightly wavy hair, a sharp face, and the walk of a ballet dancer – confident placement of every movement. Despite her slim build, she had a look of power in the way she conducted herself. "What am I letting myself in for?" he vaguely wondered, then decided to go with the flow.

There was some sort of conversation on the way. He had vague recollections of snatches. He was a computer science student, with some physics and calculus. Oh really? She was doing environmental economics, with some political science. Her family lived way out of the city – he didn't pick up where, but somewhere in the sticks. Commuting was such a bore, and not so green anyway, hence the room in a student house near enough to walk.

"Really? I ride my bike to campus. Except in foul weather – I sometimes take trains and buses."

"So you really are a greenie then?"

"Not really. Don't get on with my Dad. He offered to buy me a car, and I told him to piss off."

"That's a bit inconsistent, isn't it? You live in his house but won't let him buy you a car?"

"Yeah, well ... can't be perfect. I have an expensive habit and rent would eat into it."

"Oh?" She looked surprised – he didn't look like he was on anything.

"Computer games."

"You poor dear. We *will* have to work on you."

His head hardly clearer from the walk, he was ushered inside. The place smelt herbal, an unfamiliar odour. Someone else was there.

"What did you drag home this time?"

"Bill, don't be such a prat. This is uh Martin. I need to educate him a bit on being less of a nerd."

"I am not a lot of a nerd," he somewhat idiotically protested.

"Geek then. Split a spliff?" Bill had a piece of paper twisted on both ends to form a crude tube, one end smouldering.

"Sorry, don't smoke," Martin said, automatically.

"Idiot." Julie was laughing at him. "This isn't *tobacco*." She grabbed it and took a drag and passed it to Martin.

He eyed it out. "What's this? Smells like a burning weed." Or was that the vaguely herbal odour about the place?

"Now you're getting it." Julie was suddenly very merry.

What the hell, he thought and took a deep drag – and did a passable impression of coughing his lungs out.

The others just about collapsed laughing.

When he recovered, he said, "OK, that's lesson one. I hope it won't be in the exam." They seemed to like that. "What's lesson two?"

"Hmmm. I think a little *one on one tutoring* may be called for." Julie pushed Bill aside and led Martin to her room.

About half an hour later, just when things were getting interesting, he felt an uncontrollable urge to throw up, and rushed outside, hurling copiously into the garden.

She followed him, still giggly from the joint.

"You idiot. Dope's supposed to be anti-nauseal."

"It's not ... my fault ... if I'm different," he gasped.

"Well, I like different. But preferably not smelling of barf different. Let's clean you up."

*   *   *   *   *

"So that's how it started. Distractions. I had never been in love before. It really hit me hard. My whole life revolved around her. I still passed exams and so on, but the gaming drive was gone. I never finished that super-cool game – even though I pretty much had the whole thing worked out. It wasn't her thing, and she was my thing. She really changed me. I thought at the time for the better, but now I don't know. It was all so intense. I would have jumped through fire for her, hell, I would have taken a bullet for her.

"But I really don't know what this all was to her. There were times when she really seemed to care about me, other times it was as if I didn't exist. I would find she had organized going to a movie with her buddies, and everyone had gone when I looked for her.

"Things kind of went up and down until we graduated, then the trouble really started.

"The IT bust had happened and suddenly there were very few computer science jobs and because I hadn't gone anywhere with my game, I had nothing to get into the front door of the games business. I didn't have straight As so I couldn't get into a good overseas PhD program. Anyway I would have had to do another year. Science is a three-year degree at home . . . we have another one-year thing called Honours to get up to PhD entry standard, and I just didn't have the motivation to go on.

"She told me it was no biggie, she loved me for what I was, then managed to get a job without really trying. Strangely, in marketing or some such, but she ended up working long hours making ads – stuff for TV. Totally commercial, none of that greenie stuff – and she wasn't a plain ordinary greenie. She was into something called the green left, something like that – I went to some meetings but didn't get too much into it. I mean, these guys thought everyone was a sellout, even the mainstream Greens, who seemed a pretty principled bunch to me. I went to a few speeches by their leader, Bob Brown, and he seemed to me a whole bunch more genuine than anyone in the big parties – but anyway . . .

"It started to get harder and harder to get time with her: work kept her late, she had to look after a sick relative, she had to shampoo the cat . . . One weekend, I went around to her place, and she had some strange guy around. Seemed just like a friend, but she treated me as if I was a delivery person, or something. I didn't stay.

"Eventually, even with a whole lot of denial, I knew it was over. It was pretty hard. I was such a geek, she rescued me from that, then just dropped me cold for no reason. I used to think it was because I was such a loser – no job, no plan for my life. But maybe it was just that I took the whole thing much more seriously than she did. She rescued her little geek then let him loose on the world, like a bird with a broken wing."

He looked his new friend in the eye for the first time, nervously. "Every time I wake up, even now, the first thing I think of is her. Do you know what's the second thing I think of?"

"No," she said softly.

"I wish I could forget."

He suddenly realised his face was wet, the pent-up emotion of five years running freely out of his eyes.

"I'm sorry. It's just ... I've never been able to talk about this before. I'm making a real fool of myself, and I don't even know you."

"Oh no you're not. Anyone who saw that interview would know you are no fool. You turned a Harvard professor into jello. You know, I read somewhere that emotion is a strong hook for memory. It's normal to find it hard to forget something that hurt you deeply."

He looked for something to dry his face; a corner of a sleeve was the best he could do. Dripping on the keyboard wouldn't do.

"Allow me." She found a Kleenex.

"Speaking of rescuing injured birds, it looks like she let you loose without teaching you to feed yourself."

"Not really. I actually quite like cooking. That was one of the good things we did together. But I lost interest in food a bit when I had no one to share with."

"And you're on that low budget. I am too, but I suspect I'm less broke than you. How about I take you out to something not too extravagant, like the Union Oyster House? We can get there easily on the *T*. Food's good for the price, and they have a good selection of beer. Everyone from out West talks about the place after they visit Boston."

He was thinking of a way of protesting, then realized he didn't want to. He smiled. "That would be great. Let me put my stuff in my locker and take a shower – should we meet back here in half an hour? Just don't give me any dope, unless you have a throw-up bag." Therapy? This felt way better than sticking his face in dry academic papers and hiding behind a camera.

*   *   *   *   *

The Union Oyster House, the oldest restaurant in Boston, is also the oldest restaurant in continued operation in the US – yet an unpretentious eating and drinking hole in a convenient location, with a moderately-priced menu.

The lower level has a raw bar, where you can order freshly-shucked shellfish of various kinds. It also has table seating, and a more restaurant-like upstairs. There's also a drinks bar – all surrounded by warm wood.

When they got there, the lower level was full, but they managed to get a table upstairs after a short wait. A fair fraction of the people there were family groups. The place had a happy, homey atmosphere.

"Have you ever had clam chowder?" He was reading the menu, wondering whether scrod was a local name for something otherwise known in the rest of the world. He looked up at the interruption.

"No, I haven't. It's not big in Australia."

"Well, you need one of those to start, to get some local taste. And speaking of local taste, have you had a Samuel Adams?" He guessed she meant a beer.

"No – I've heard bad things about American beer, so I didn't look at any."

"This is different: not quite as mass market, made in a more European style."

"OK, then. I'll try clam chowder and a Sam Adams."

"That won't be enough on its own."

"Then the smaller size of chowder. What exactly is 'scrod'?"

"I think it's sort of generically anything that could be passed off as cod."

"Oh. Presumably OK though . . . it's been years since I had scallops."

"How about we get both and share?"

"Great idea," he agreed. Sharing sounded very pally.

They ordered, and two cups of steaming chowder with bags of oyster crackers on the side appeared quickly, and, soon after, a pair of Samuel Adams bottles with frosted glasses.

Martin took a moment to realize that the oyster crackers played the role of croutons. He tasted one – rather like what was sold at home as a water biscuit. Not much taste, but a good crunch. He emptied the rest into his chowder, following Angie's move.

"Pretty good. I should get this again before I leave."

"Speaking of which, exactly how much longer are you here?"

"After the weekend, I'm going to New York to see some NASA people to add to the mainstream, then back home. I suddenly wish I was staying a bit longer."

She reached out for his free hand and played with his fingers. "I'm heading back to Berkeley on Monday – bit of a rush for first day of semester Tuesday, but weekend flights were full. Did you have any plans for the weekend?"

"I did, but I can think of a better way to use the time. If you aren't too busy to help a lonely guy have some fun." He gave her an engaging grin, his plans for reading everything he could on the NASA view of ice melts suddenly seeming less important.

She laughed. "You're not so bad. I'd say the wing is well and truly healed. Of course I have nothing better to do. Boston is a great place to lurk around, and it will be so much more fun with company. And this is the best time of year – bright leaves starting to show, starting to cool off."

The scrod and scallops arrived, and they had to make space on the table.

The food was good though not spectacular, but he hardly noticed. He still had a lot of barriers up, but this felt distinctly like falling in love again.

They left the crowded restaurant after he made a muted protest at her insistence on paying, silenced by – "You forget,

I took *you* out."

He put a tentative arm around her, and she pulled him closer.

They didn't pay much attention to where they were going, and ended up facing a darkened Faneuil Hall market building. She said, "This will be a place to visit, but not today. I think the *T* stop is the other way."

They turned around to face the opposite direction, a movement which naturally segued into a long, lingering kiss.

"Come on, we're wasting time." Angie tugged his arm, and towed him towards the City Hall neighbourhood. "There's a *T* stop here somewhere, I think – I remember passing these buildings on the way out."

Half an hour later, they were back at the Parrot. There was someone at the front desk, looking about ready to pack up for the night. She pushed Martin away and had an animated, not totally audible conversation with the person at the desk. Eventually, she turned to Martin, and said, "Let's see your key." She turned to the desk and showed the number on the tag. Some nods were exchanged.

Finally she turned back to Martin, who was waiting quizzically. "I hope you don't mind, but I've been haggling to get a private room for us to share for the same price as our shared-room dorm beds. You don't mind, do you?"

He didn't.

*   *   *   *   *

Saturday evening. Almost the end of another perfect late summer day in Boston.

They were on a boat on the Charles, in the glide path of planes landing at Logan. Although the air was mostly clear, there was a hint of fog on the river, just enough to make it look as if the planes were appearing out of nowhere, almost close enough to touch, as they majestically lowered themselves to the runway, the end of which faced their boat as they headed out to the harbour.

"A great idea, doing a harbour tour," he said. "I'm not much of a tourist. I usually go somewhere to *do* something." They were standing at the rail of the boat, bodies pressed close as if to absorb the gentle rocking motion.

The memory of the night together still lingered, something special after so long on his own.

"You were right. Faneuil Hall was so much nicer in daylight. And open."

She laughed. "And it was great walking around the city with friendly company. Berkeley is also a great walking city – lots of little coffee shops, student hang-outs ... but I suppose you had that at home? When you were a student?"

"Not really. The University of Queensland is pretty boring. It's near enough to the city centre that it doesn't have to be self-contained, but far enough that it doesn't have non-student crowds to support a wide range of student hang-outs. And a lot of my fellow students couldn't wait to get out of the place when they didn't have to be there. I never understood it myself. You pay good money to be a student – even if someone else is paying the bills, you could be out making money doing something that needs no skills, like flipping burgers, or politics.

"I wanted to have fun while I was there. The only thing

was, my definition of fun kept changing. I wanted to build the greatest computer games at first, then there was my other . . . distraction."

"Are you having fun now?" Her impish grin looked like a kissing opportunity.

"Did that answer your question?"

It was getting cooler on deck, so they held each other closer for warmth, as the boat turned, and started to head back.

"Where next?" he whispered.

She snuggled closer. "I hear the John Hancock building has great views if you like city lights."

"I like."

By the time they found their way there, it was getting dark. But there was a problem – the observation deck was closed. The concierge told them, "Sorry, I get this several times a day. It was closed after September 11. The highest point you can still get to is the Prudential Building." Noting their disappointment, he quickly added, "It's only a few blocks away." He gave them directions.

It turned out that the view site was a restaurant, rather more expensive than their class of budget. The view was pretty good, so they lingered over inspecting the menu for a while, and walked out.

"That was a pretty good view for the price." Martin played with her hair on the way down from the fifty-second floor.

She laughed. "But I suppose we should get some food. How about something really upmarket like Chez Subway?"

"Only if you let me pay this time. And let me get a bottle

of wine if we can find a place selling something Australian."

They headed back to the hostel with their purchases.

"Now, the next trick will be to find some wine glasses," he said as they walked into the Parrot.

"And a corkscrew."

"No problem there." He revealed the top of the bottle. "Screwcap. They're using them on even the fancier labels in Australia. More practical, less likely to leak and let in air, or go mouldy."

"But no romance."

"That, we supply."

The crude kitchen table, the passably good Cabernet Sauvignon-Shiraz blend, the fast-food sandwiches, all blurred into insignificance.

* * * * *

Sunday morning. A lazy wake up.

"What are you thinking about?" She saw him gazing into the distance, to the extent that was possible in a small room with tiny windows.

"You."

She smiled. Neither of them mentioned what his waking thought used to be.

"This is our last day together. Tomorrow, I'll be in New York, you'll be in Berkeley."

"Last day ... you make it sound so final. You said you were only making a pilot – when you do the full thing, you have to visit Berkeley and Stanford. There's the California Climate Change Center for a start. You don't want to give the flakes too much time. And there are some really great people at Stanford."

"If."

"No. The stuff I've seen is great. It just needs a bit of editing."

"Yeah. That part's not going to be fun. Stitching together a few clips with entry level free stuff on my Mac is not so bad, but it's going to be hell doing serious editing. Could take weeks."

"Can't you go back to the production company who's advancing you the money – use their equipment?"

"Everything is outsourced these days. They don't have editing facilities. I need something good to show off to get more money out of *them* let alone to sell the thing."

"Now wait a minute. You mentioned someone who is in advertising – does she have access to high-end video editing equipment?"

"You mean . . . Julie?"

"Sorry, I was hoping to train you out of thinking of her first thing, now I bring it up. But if she really is such a greenie at heart, even if she's sold out a bit, she should be excited to see what you're doing. I can't believe she really had no feeling for you. Did it occur to you that *you* may have outgrown her, and she ran away before you dumped her?"

"No. Never."

"Look at you now, doing this green thing. Don't you think you are trying to do something she'll notice?"

"I don't know. I mean I started this whole thing as maybe the so-called sceptics are right – but no one has put their case based on the science. Not convincingly."

"OK then, to get back at her."

"No, no. I hate being lied to. My whole life, someone has been lying to me. At an early age, my parents separated, before I was old enough to know. My dad kept me, and my mother was this distant figure. They had some story about it but I kept finding inconsistent details. They just changed the story. I mean, something like half of all marriages end in divorce. Why couldn't they just tell me the straight story?

"Then there was Julie. I was ready to trust someone, and she also lied to me. After we graduated, it was one story after another. I don't know why I didn't see through it sooner."

"Shampooing the cat?"

"Well, I made that one up but it started to feel like that. No. No. This whole thing was about my own little fight for the truth, whatever it was. That's why Schoor pissed me off so royally. He is a lying git," he concluded through clenched teeth.

"Jeez. I'm sorry. Can I promise never to lie to you? At least to try pretty hard not to?"

"I'd like that, but I'm not as fragile as I used to be. The odd little fib is sort of normal. Just not the life-changing stuff – but let's not talk about that now." He looked serious. "We have more important business."

"Oh?" she asked with her eyebrows.

"Snuggling, for a start."

After some time, they surfaced, and Angie said, "But still, if she has the video editing stuff, ask her. If she's horrible you can always call me for comfort."

"Well, all right. I'll give it a try." He didn't sound confi-dent, and changed the subject. "Is there anything else you'd like to do in Boston?"

"How about some art and museums? There's the Museum of Fine Art, and you may like the science museum."

He pulled his computer out from under the bed, and did a search. "How energetic do you feel today?"

"Didn't I already demonstrate how energetic I feel?"

"And how. Look at this." He showed her a google map. "There's a huge cluster of museums just across the river from here, just one or two *k*s –" she looked puzzled for an instant – "sorry, about a mile – from here."

"I know what a '*k*' is. Not all Americans are ignorant dolts. But yes, it looks good outside, and walking is always good. Especially as I haven't been getting my usual exercise for a couple of days."

"What's that?"

"I like to swim. I don't feel the day's right unless I've done at least 10 laps." She squeezed a biceps, as if looking for flab.

"Well, I don't think you want to swim in the Charles, so we'll have to walk."

*   *   *   *   *

They walked out of the Museum of Fine Art.

"That was pretty good. Though I was surprised how much you have to pay. In Brisbane, most of our art galleries are free."

"I've always wanted to visit Australia."

"There now, you have a reason."

"No, two. I already feel I'm going to miss you."

They were walking along the river towards Boston University, when Martin looked between the buildings, away from the river.

"Hey, look at that. The green line thinks it's a tram."

They walked away from the river towards the train. "Didn't you know that?" She explained: "The green line has three different modes of operation. It's part subway, part suburban train, part tram.

"It's the only part of the *T* system with two different fares. You pay the same as any other train in the city, even if you use it to go out to the outer suburbs. But you pay extra if you board it in the outer suburbs, heading to the city."

"Weird."

"But nice, it can take us back to the red line, and home," Angie added.

"Home has a nice sound to it."

"Where exactly do you live then?" she asked.

"Not exactly in a hovel. I rent a place in the slightly less fashionable side of Brisbane. Not too far from the city centre, but shabby enough not to be expensive. It's actually better than it sounds. I don't have to walk far to get to passable restaurants, go to a movie, or catch a bus or ferry."

On the train, she said, "What exactly is this IT bust thing you keep talking about?"

"A lot of people think the dot com thing was a bubble, but I don't think it was that. The big players are still there and you can buy more on-line than ever. The Y2K scare had a lot of tech-ignorant managers rattled, and they spent three years worth of budget in one year. That created a short-term boom, then there was no money for new development for a couple of years. The year I graduated, we had a huge class and a lot of us battled to find jobs. I hear the classes are tiny now."

She shook her head. "That's crazy. People haven't stopped using computers. Everyone I asked about doing a PhD told me I should do some computer science – you need it in any of the new bioinformatics stuff."

"What computer science have you done?"

"Just the basics, programming, algorithms, some discrete math."

He thought for a bit. "Is that what you want to get into, bioinformatics?"

"Yes. I first got into biology with just a general idea that I wanted to do something that would help people, make the world a better place. I thought of med school, but I wasn't sure if I was cut out for that, so I decided to go for general bio first, and if that worked out, I could go into something more applied, like bio-ecology. Then we had a prof from Stanford visit and give a talk about gene sequencing. He started with a quote from Rutherford, the physics Nobel –"

"I know who Rutherford was: *all science is either physics or stamp collecting.*"

"Exactly. Anyway this prof said that biology used to be stamp collecting – arbitrary classification – now for the first time, we were developing a real science around it, something with a theoretical model you could test. And computer science was a big part of that."

"So why MIT? Isn't Stanford pretty good too? And Berkeley has to be good in this too – they are both so big in computer science, and I'm sure Stanford is big in biotech."

"Well, I have to admit . . . I really love my family but I

want to get away. My parents are both professors and it would be nice to feel I was going home away from professors. Not that professors are evil or anything, and I love these two dearly. My mom is the best mom in the world, and my dad is wonderful, but sometimes I feel smothered by all that love – do you know what I mean?"

"Sort of." His answer sounded like, "I wish I did." The conversation dried up for a bit.

"This is where we change for the red line." She pulled him up off his seat.

They had to wait a few minutes for the train.

"So, only MIT? I mean, even hotshots like you get turned down."

"I'm not silly. I applied to a few other places but MIT has the people who excite me most. I'd have to think twice about any of the others."

A silence followed. "Maybe I can give you some hints about what CS to get into . . . help you refine your list of extra courses." Martin told her about his experience at UNSW.

She looked impressed. "You showed up a room full of PhDs, just like that?"

"Well, it was a *little* room."

"You're so modest. Come on. Bask."

"Yeah," he wasn't used to this and lost his train of thought for a minute.

"Hey, that's enough basking. Tell me what you did. I want to pick up some hints."

"Not basking, stupefied . . . You want to take up climate modeling now?"

"No – but bioinformatics is heavy lifting too. Why did that stupefy you?"

"Not that. I'm used to modesty."

"OK, so what *did* you do? Impress me with the details."

"How much computer architecture did you study?"

"None really, just an entry-level course."

The train arrived, but the conversation hardly missed a beat. This time, they had to stand.

"OK, so your average joe who goes out and buys a computer looks at how many gigaherz the processor does. Up to a point, that's good, but for big computes, you have to look a lot more at how you use memory. Most of the memory is pretty slow compared with the processor, and you have to organize everything so you are in the fastest part of the memory as much as you can."

"So why didn't those geniuses in New South Wales get it right?"

"First, most people do this stuff in separate courses and don't put it all together. In my gaming days, I wanted to understand everything, and did put it all together. So: big hint number one for you." She nodded as he continued: "Get the big picture.

"Second, it's a whole lot more complicated with multiple processors. If you get something into the fastest memory on one processor, and another processor rips it out, the two end up fighting over it and not getting any work done."

"You're talking about the caches, I take it," she said dryly. He looked contrite; note for the future: this is someone you never talk down to.

"Sorry, you said you'd done virtually nothing ... but I forgot ... Berkeley ... one of the great centres of computer systems. I learnt a lot from my courses, but I practically spent half my life on Berkeley web sites.

"So – a few more things to worry about. Cache optimizations like eliminating false sharing, and blocking. You can thrash the TLB if you jump around the address space too much. Spinlocks are terrible in shared-memory systems. Disks like to be accessed in big chunks at a time because the latency is so much slower than the peak transfer rate ..."

The flow of words was getting lost in train noises.

"OK, OK. You've convinced me. I probably could use a few more courses. Maybe I don't know one or two of those things. 'TLB' sounds a tad unfamiliar. Some computer architecture, you say?"

"You said something about algorithms before. How good is your coding?"

"Let's just say I scare my gang, and some of them are CS majors."

"Wow. Do you bask?"

"Only when they're really terrified." She tried to look modest.

He took a minute to digest this. This goddess can't be impressed with me. But to hell with it. Seize the moment. What was it in that movie? Dead poets or something? ... Oh yeah: *carpe diem*. "So ... operating systems ... and maybe networking ..."

"Networking? What does that have to do with big computes?"

"Building really gigantic computers out of lots of little ones. Check out the research at Berkeley. Good stuff happening there."

"And operating systems?"

"Understanding the plumbing underneath. Paging, for instance, like the TLB I mentioned – it stores recent page translations."

"OK, cool. That's pretty much my semester planned. More CS, so I'll have a think about what of the other stuff to keep."

They arrived at the hostel.

"Let's dump our stuff, and get some food." Angie had the room key out, and he followed.

"Yeah. Did you see the food store a block or two back?" He pointed back up the road, and she nodded. "How about we cook something? They have the basics in the kitchen here, and I have a lot of experience working with the basics."

"OK. This time, let me get a California wine. There's a liquor store right next to the food store."

On their way out, they looked into the kitchen. A couple of other inmates were already there, checking out the equipment. They introduced themselves: Pete and Inga from Sweden. Angie explained their intentions.

Inga said, "We've got some spaghetti and olive oil. Too much for two, and we're only here one more night."

"We can add to that. I can show you fusion cuisine, Truscott style – if you don't mind me taking charge."

The Swedes nodded.

"Not a problem," said Angie. "I'll see what I can make for dessert."

"Like what?" Martin raised an eyebrow.

"Surprise."

All four of them went out to the shop. The Swedes agreed to get some parmesan, and split off to head for the cheese area.

"Right, no peeking," Angie said, and headed off to the fruit area.

Martin shrugged, and picked up a basket. He found a can of Moroccan sardines, a bag of raw macadamias, and then headed off to the vegetables, where he contemplated various options, hefting an eggplant, artichoke – no, too much effort. "Yes!" He found a very fresh bunch of broccolini.

Angie, by now outside the checkout, looked his way. "Found what you wanted?"

"Found something good – I only had a general plan for this part." He also picked out a shallot.

As he paid, Angie said, "You really had some quality time with those vegetables."

He grinned. "I'm green. I like plants. What did you get? That looks like fixings for a gigantic dessert."

Angie hid her bags from him.

"OK, OK – I promise not to peek."

They found the Swedes outside the store. They had bought an impressive looking chunk of Reggiano. On their way back to the hostel they swapped stories from their day out.

Back at the Parrot, Martin issued instructions for the pasta. "Hey, we know how to cook," Inga complained.

"Sorry, I didn't cook with friends for a long time." Martin contritely focused on his side of the action. He set a pot of water to boil, and looked for some salt. There was a bag in the back of a cupboard, along with a pepper grinder. Hoping these weren't personal treasures, he hauled them out. Judging from the dust on them – no. He added a good mound of salt to the water.

Angie looked up at that moment. "Hey. What's with the high sodium?"

He laughed. "Don't worry, that's just to make it boil hotter. I'll wash the salt off afterwards." She looked doubtful, but carried on at her end of the kitchen, whisking and stirring. She brought a bowl over to the stove, and found another pot. She added water and set it to heat next to the pasta. When it was near boiling, she added the bowl to it, double-boiler style and started whisking vigorously.

Meanwhile, Martin's water was boiling, and he dumped the broccolini in and closed the lid down tightly. They were running out of pots, so he peeled and chopped the shallot, while waiting for the water to reboil. Almost as soon as it started boiling again, he inspected the vegetables and, happy with the colour, pulled them off the heat, emptied out the water and rinsed them off. "See! Not salty at all." He chopped off pieces for the other three to taste.

Angie showed appreciation. "Wow. I didn't know they could taste like that."

"Just you wait."

"Uh oh. I nearly forgot mine." She went back to the whisking. "Just a few more minutes . . . "

"Hey, guys – how's the spaghetti?"

"About two minutes," Inga replied.

"Where's this olive oil then?"

Pete passed it over – a small bottle but more than half full. Martin dried off the pot he'd just used, put it back on the heat, added a good dose of olive oil and tossed in the shallots. They hissed indignantly, so he lidded them to shut them up. In a few seconds, he added some macadamias, then opened the sardine can, drained it, and broke the sardines in coarsely. Finally, he tossed in the cooked broccolini, pulling them apart a bit. He closed it up again, and yelled, "When the pasta's done, pass it here."

A minute or so later, he poured the drained pasta over his ingredients, and looked around to see what Angie was doing. She'd stopped work, her bowl off the hot water. "Mine will need a little finishing, but let's have the pasta first."

The Swedes found a grater, and between the four of them, they found plates, cracked some pepper, grated parmesan and plated the pasta.

"Wow, looks great," said Angie.

"Yeah," the Swedes chorused.

"Oh, crap." Everyone looked at Martin. "We forgot the wine."

"We? You." Angie pointed at him. There was a bottle on her side of the kitchen. She brought it over and opened it. "Hope you guys like a nice Pinot Noir. Sorry it has an old-fashioned cork."

No one minded.

She rooted around for a corkscrew, and found one – Italian waiter style. A bit rickety, but it worked. You can't spoil a good design.

Martin was puzzled. "When did you get that?"

"Marty boy, while you were communing with vegetables, I remembered the liquor store next door, and sneaked out to get this. And these." She lifted up a box and produced four extravagantly proportioned glasses.

"We can't drink this out of hostel tumblers. Besides, when I move out to my own place, I'm going to need some glasses."

Martin hefted one, marveling at the subtle shape. "Classy."

"The best. Riedel," Angie added solemnly.

The Swedes looked impressed. Martin nodded encouragingly. Better look it up . . .

The wine *was* good. So was the food.

"Where did you learn to cook like that?" Pete asked.

Angie looked at the Swedes. "I only met this guy two days ago, and he's full of surprises."

Martin explained. "Supper club. Last year or so of uni, Julie – my girlfriend then –" the explanation for the Swedes – "had this idea that we should take it in turns cooking. Give ourselves a treat while learning the tricks of the trade, cause we couldn't afford good restaurants. It turned into a sort of competition."

"Yeah," said Pete, "keep surprising us too."

They all laughed.

Martin added, "We also sort of competed to be greener in cooking style – vegetarian ideally, but I kind of justified fishes to myself – not quite the energy waste of growing a cow."

"What do you call that?" Angie asked. "Vegaquarian?" They all found that pretty funny.

The pasta out of the way, Angie said, "Let's finish the wine. The dessert won't go with it."

There wasn't much left so in a few minutes, she went back to the far end of the kitchen, and whipping and mixing continued. Then she returned to the table with four mugs.

"Ta-dah! Sorry about the presentation."

"What's this?" asked Martin.

"Taste."

He did. "Your turn to bask." The Swedes looked puzzled. He explained basking theory briefly. They tasted, and agreed.

"So," he said. "I would guess an animal of the moose family." The Swedes looked puzzled, and vaguely horrified.

Angie laughed. "That's M-O-U-S-S-E not M-O-O-S-E." The Swedes looked relieved.

"Yup," she added. "Lime mousse made with good natural ingredients. Lime juice, sugar, butter and egg yolk. All that whisking over a hot stove was to make a custard of butter and egg yolk. You add cream to cut the fat."

Martin pulled his shirt tight to see if he could still see his ribs.

*   *   *   *   *

It was nearly eleven by the time the party broke up. Martin said, "We're both leaving tomorrow. Great to meet you guys – in case we don't see you tomorrow." They walked upstairs together.

Pete said as they split off to their end of the corridor, "Are you sure you guys only met two days ago?"

They said goodbyes, then Martin had her to himself again.

They took separate showers, and met in the corridor. He stared into her eyes, then they opened the door and went to bed.

"Lucky we don't have early flights," Martin said.

"Yeah, I can see you off, then lurk a bit first. Not the same without you," she added.

They lay together in silence for a bit. Then, he said, "You know, what Pete said . . . "

"Yes. It really doesn't feel like only two days," she agreed.

He went on: "I can't believe this is our last night together."

Angie put a finger on his lips. "Not. You are going to sell your movie concept and visit me in California, or I will hunt you down in Australia. This is *not* . . . ."

". . . our last day. Did I tell you yet that I love you?" he finished.

She didn't reply verbally.

As he fell asleep, his last thought was her word, "*Not* . . . "

7 NASA

GEOFF BLUNT sometimes felt he'd be better off chasing his own tail.

Between running mega-experiments with all that entailed – so many things that could screw up, so many bureaucrats to deal with before you could stop pushing pencils and do something real, traveling twice around the world to stitch up collaborations that NASA would rather you did without – he was writing letters to the press, writing articles for his blog, correcting misconceptions in comments on his articles on his blog ... it was a wonder how he ever got anything useful done.

Yet he had a satisfactory list of publications, some in pretty respectable places, and occasionally, just occasionally, someone would listen. And of course, he wasn't alone. Thousands were working on the problem. But there was the other side. The deniers. Everyone on the inside knew that apart from one or two who were either masters at self-delusion, or really had something but couldn't explain it to other scientists, much of the "denial" camp as he preferred to label them were either industry shills or attention-seekers, past their prime and

no longer able to do useful work. But the press loves controversy, and building the whole climate change thing up as the
valiant deniers versus the well-funded, well-oiled propaganda
machine of the "mainstream" science makes for a much better story than the reality. The reality that it was in fact the denial camp that had the well-oiled propaganda machine without much substance, rather than the other way around. And
the more you argued it, the harder it became, because the response was, "You would say that." If only they would actually
do their jobs and figure out that the shills were not actually doing real work, and the people like him were actually putting
in the hard grind.

Now there was one of them at his door. One who sounded
suspiciously like *Son of Swindle* – despite some good introductions. But duty called. He walked to the door.

"Truscott, Martin Truscott," said the visitor.

"Right, I was expecting you – anyway I don't get a lot of
visitors with video cameras."

Truscott glanced around the office, piled high with papers, four computer screens around a cluttered desk. Blunt
was almost as skinny as Truscott, but a good bit taller, with
a stubbly beard, ending in a slightly more convincing patch
on the chin. Hair thinning, he looked older than he probably
was.

"Mind if I get a shot of you at your computer, for context?" Truscott asked.

"OK, sure." Blunt grimaced and did working scientist im-
pressions for a few seconds. "Are we done yet?"

"I suppose . . . should do for now. Where can we talk? I
want to get the details straight before going to interview. I've

read a lot on your blog ..." Blunt raised an eyebrow at this "... but accuracy is important to me. More than to some of my colleagues."

"Depends how far you want to go. If you want detail, we should stay within WiFi range so I can look up papers for you." Blunt threw that at him as a challenge.

"Great, I want detail. Not sure how much your office told you. I was not impressed with what Durkin did. I want to spike this whole debate by getting to the facts, at least as far as we can now."

Blunt pulled his notebook out of its network connector and charger. "I thought Al Gore did that already."

"Yes, but no one has challenged the 'sceptic' bunch directly. Durkin treated them as gospel. Gore more or less ignored them."

"Well, OK. Let's go to the coffee shop across the road. They should be quiet now, and they have a WiFi hot spot there." Blunt hadn't revised his opinion of the potential for wasted time, but it did feel like time for a coffee.

They walked out of the building and across the road.

*   *   *   *   *

After they ordered and perched at a table in a quiet corner, Martin pulled out his computer, but left it closed. Blunt echoed his move, then turned to face his visitor. "OK, so where would you like to start?"

"Well, let's start with areas where the science is still unclear, areas of unknown ..."

"That's a pretty wide field. You know, when we say the science is settled, it doesn't mean we know everything. There are a lot of gaps, a lot of areas of correction. If you look at the IPCC reports, they cover a range of scenarios ... "

"Business as usual etc."

"With a whole range of variants on how much more or less CO_2 is spewed out over time. But even in a given scenario, the modeling has a range of values to take into account inaccuracies in the model and measurements. So you get a range of temperatures, a range of sea level rises, and so on."

"Right, I understand that. What I mean is some of the uncertainties people talk about, like whether the Antarctic or Greenland ice caps are melting. Some of the stuff I've seen claims that the ice is actually growing, rather than shrinking." Martin started to feel a bit uneasy about his neglected weekend reading. Wilkinson had told him exactly what to look at and he hadn't ...

"If you look at the latest IPCC sea level estimates, you will see that they actually took out the ice melting calculation because of the uncertainty. But if you look at sea level rises, something is clearly happening to the ice – there's no other explanation. The IPCC models sea level mainly as thermal expansion. Their attitude is that it's 'prudent' to leave out something they can't fully explain, even if it could be a major risk factor."

"Major?" Martin started to pay serious attention. "In what way, I mean is there something big the IPCC is missing?" This would be less of a surprise if he'd done his usually meticulous homework. He tried to remember what Wilkinson had told him ... "So the IPCC had scaled *down* their estimates of sea level rise ... and that's only because they took out a factor they didn't understand? Wilkinson at University of New

South Wales told me something about this ... but I was thinking this wouldn't cause a major upward revision, otherwise more people would be talking about it."

"Well ... some of us are tired of being called alarmists ... but the West Antarctic Ice Sheet is looking increasingly less stable than we all thought a few years ago. You could say that our previous understanding is fracturing, but not in a positive way. All the indications are that WAIS could go a whole lot faster than expected."

"How big a deal is that?"

"Big. The IPCC is talking maybe half a metre of sea level rise, at the upper end of the range – actually 59 cm. If the WAIS goes, that's more like five metres."

"Wasn't some of this in Al Gore's movie?"

"Yes. But the science of it is only now starting to get clear. He only speculated on what various sea level rises would mean in various scenarios. We didn't know at the time whether any of that was likely."

"And now?"

"We still can't be sure, but let me get you some of the latest NASA work on WAIS, and some that the Brits have been working on for bed time reading, so you can see for yourself. Some of us are pretty convinced that a multi-metre sea level rise over the next century is almost certain if we stick to business as usual – let alone accelerated CO_2 output." He glanced at his watch. "Sorry, got things I have to do, but this shouldn't take a minute."

As Blunt was pulling in various papers, he looked up. "Can I email you the links? Or should I print these for you at the office?"

Martin thought fast. Low budget, fighting with hostel facilities ... "You know, I'd like to look at this stuff as soon as
possible ... I feel like an idiot, planned to read your papers
on the weekend but got sidetracked ... so if you could print
them, and point out some of the main issues to follow up, that
would be great, then I can start my reading on the subway as
soon as I leave ... should be in good shape for the interview
tomorrow."

"Oh yeah, that." Blunt grimaced. "Sorry, there's been
some really stupid press on this subject."

"Well, I am not your average journo. I have a computer
science degree, with some calculus and physics on the side.
After the Y2K bomb, I switched over to making TV science
shows. Maybe I am not a climate scientist, but I think I do
understand a bit more about science than most of the press."

Blunt looked noncommittal, as he sent several papers to
the printer, then closed his computer.

As they walked back to the NASA building, Blunt added,
"In case you think this couldn't happen, around fourteen thousand years ago, sea levels increased about twenty metres over
four centuries – about a metre per two decades. And the forcing then was not as strong as it is today."

Blunt took him to a facilities room, and handed over a
sheaf of papers. "Happy reading. Start with the *New Scientist*
stuff – main points covered; less depth. Same time tomorrow?"

*   *   *   *   *

The Goddard Institute for Space Studies, or GISS to its
friends, sits above the diner made famous by *Seinfeld*. Part

of the urban Columbia University campus, it is one of the hubs of the US climate change effort, and one of the main centres of what some like to think of as climate alarmism. It is also the home of one of the main contributors to `real-climate.org` – not Blunt, one Ian Grant.

Up till now, Martin had been eclectic in his `realclimate` reading – NASA, US universities, UK and European climatologists. Oceanographers, physicists, geochemists. He didn't care much who wrote the article, or who posted follow-ups. The duel between the two sides was there, and understanding a bit of the science helped. But the good people who took the time to write also explained a lot for the uninitiated. It was all good reading. But after getting back to his hovel with his reading, he decided to to set up a meeting with Grant while he had the opportunity. Grant seemed to be keen on taking on the sceptics. With Martin's growing doubts and uncertainties – yet increasing knowledge – he felt he had to talk to Grant while he had the opportunity.

So that night, he composed an email.

```
You don't know me but I read realclimate a
lot.  I am visiting Geoff Blunt to interview
him for a TV show I'm working on, about climate
change.  I wonder if you'll have a few minutes
for a short chat tomorrow to clear up some
issues.  I'll see Geoff at 11:00, at most
for an hour.  The rest of my day is pretty
open.  If you could spare a few minutes, I'd
really appreciate that.
```

"Always pays to be polite," he thought, sending it off.

*   *   *   *   *

"Goddam it, Geoff." Grant was not in a happy mood. A bad day at a congressional hearing, a stupid article in the *Wall Street Journal* ... it would all be spin to counter on `realclimate`, more time away from work. "Who the hell is this Truscott? Because he is seeing you, I felt compelled to offer to see him too. Don't we have enough to do without time sinks like this? He's going to suck everything he needs out of us and ignore everything he disagrees with, and leave us with yet another counter-spin exercise."

"Calm down, Ian. For a start, I didn't tell him to get in touch with you, or tell you to accept the invitation. Secondly, I had a pretty good talk with him yesterday. He has some science background if not in climate, and seems more willing to listen than the regular bunch. Give him a few minutes, and check him out for yourself. If it looks bad, you can always make up some emergency."

Grant grunted. "I don't have to make up emergencies."

"Join the club. Should I pass him on to you when we're done?"

Grant looked resigned. "I guess. That would be about lunch time. Maybe he can take me to lunch ... "

"Don't get your hopes up. He looks pretty hungry to me – about the lowest-budget TV production you'll see outside the third world."

"Hmph." Grant liked a good feed, and looked it. Rather more rotund than Blunt, with a beard rimming his face, he was overall what some described as a well-rounded personality (that is, those who only observed the visual aspect). "Maybe

I should get him lunch instead ... bribe him to do a decent story."

They went their separate ways, Blunt to his office to catch up on some writing, Grant to his lab to oversee some measurements.

Meanwhile, Martin was exploring New York's famous street coffee culture, in the form of a piping hot coffee and a bagel with cream cheese. Cliché perhaps, but good.

He took his time. The coffee was hot, and managing a coffee, a bagel and his gear was a challenge, especially as there was nowhere obvious to sit. The coffee and bagel eventually disposed of, he found a copy of the *New York Times* and, for good measure, the *Wall Street Journal*. Martin generally did not care for the business press. Business people had an arrogance about them, as if they were the only ones who knew the way the world worked. But this was New York, and he felt the need to pick up some local colour.

He suddenly realized he was only a block away from Central Park. "Fool. I've been juggling all this stuff when I could be sitting on a park bench. The subway station before last had to be called 'Central Park North' for some reason." He found the location of the NASA building, then backtracked to the park, calculating how much time he would need to get back. "Five minutes, tops," he thought as he found a place to park.

He put his gear down next to him, maintaining contact with the backpack – uncertain whether New York's high crime reputation was totally a thing of the past – and flattened out the newspapers. The *Wall Street Journal* fell open. He was about to close it and start on the *New York Times* when his eye caught a headline on the opinion page.

He picked up the *Wall Street Journal* and read the opinion piece, his eyes narrowing as he read.

He suddenly realized he had three minutes to make his appointment. He gathered up his backpack and the papers, and rushed off to the NASA building.

He arrived, panting and looking pretty wild at Blunt's office.

"Uh oh. I was just telling Ian you're OK, and you come charging in here like a lunatic. What's up?"

"Sorry," he brandished the *Wall Street Journal*, "have you seen this crap?"

"Not only have I seen it, but I rather think Ian is going to like you more than he expects. *WSJ* is not our favourite publication. They have a long history not only of failing to get the facts straight on climate change, but of contradicting themselves. They made a big deal of the Medieval Warm period, a concept which has a fragile basis in science, as well as the Little Ice Age, for which there is just as much evidence. Then to back up their claims, they cite a paper which puts a period of unusual warmth into the fifteenth century, in the heart of the Little Ice Age. Morons. They should stick to economics, where you can change the answers every day, and no one pays attention."

"Little Ice Age and Medieval Warm Period ... the contrary crowd talk a lot about those. What's your take on them?"

"The data to support both is patchy. There's evidence of higher temperatures in different locations and different times in the 'Medieval Warm Period' – but not consistent data to show warming everywhere at once. Anecdotes like producing wine in Britain in that period are laughable – more wine is grown in Britain now than during that period. All flimsy stuff: falls apart when you take a hard look at the data."

"Right. I've just last week been at Harvard, and interviewed Schoor. He couldn't back up any of his claims when I pressed him for detail, yet they quote him in this article as a leading authority. He went ballistic when I asked about his fossil fuel funding."

"Really? I would like to have seen that."

"Well, you won't see it in this publication." He looked for a place to put the paper down, then realised he was in the doorway with nothing in reach. "I could show you the clip, if you promise to keep it under wraps."

Blunt said, "Sorry, I'm being rude." He ushered Martin into his office. "Now, where were we last time? And yes, I would like to see the clip."

"We were talking West Antarctic, and you gave me a week's worth of reading ... Which I stayed up all night reading, because I shouldn't go into an interview under-prepared." He passed a DVD over as he was talking. Blunt took it, inserted it into his computer, and, seeing there was only one file, copied it over.

"Sorry, I didn't expect you would read it all at once."

"That's OK, I am still not totally in the time zone. I am aiming to stay awake for two whole days so I can sleep properly tonight."

"Haven't you heard of melatonin?"

"Yes, but I dealt with *that* in another investigation ... I'm not sure anyone should use it ... anyway, to business. I am interested in what the latest scoop is on the ice sheets – not just what's been published. The trend is worrying. Papers in the 1980s were saying that the West Antarctic was stable and had a close to zero probability of disappearing fast. But everything more recently is more or less saying that all past theories of dynamics of big ice sheets is wrong. Things like streams in the ice moving faster than expected."

"You're in luck." Blunt was looking at one of his computer screens. "Will Wilson, one of the leading Antarctic researchers, is online. Let's Skype him and see if he's willing to chat. So it's not just me talking." He ejected the DVD and handed it back.

"Where is he?"

"In the UK. British Antarctic Survey. Isn't technology wonderful? All we need to talk to him conveniently is synching our time zones. It's mid-afternoon there so we may be lucky and find him not too busy."

He activated a call on his computer. After a few rings, Wilson's face appeared on the screen. "Hi, Geoff. To what do I owe the pleasure?"

"I have a young fella with me, one Martin Truscott from Australia, who is doing a TV show on climate change, kind of anti-*Swindle*. He wants to get the straight story on the Antarctic from the horse's mouth."

"Hmm, I don't know. Can we trust him?"

"He's right here, so you could ask him yourself." Wilson looked awkward. "Or better still, let's see if I can position my camera so you can see an interview with Schoor he just gave me. If that's OK with you?" He turned to Martin, remembering that he wasn't meant to share it.

Martin nodded. "But please – this should go no further. I don't want to spook other interviewees with the incorrect perception that I'm taking sides."

Martin sat back, thinking, this will be a good test ... we'll soon see if Angie was just being nice.

Camera in position, Blunt launched the clip. "Can you see OK?" he asked Wilson.

"More or less. It's only breaking up a little."

They watched in increasingly stunned silence. At the end, both scientists were speechless for a moment.

"Wow." Blunt broke the silence. "I hope I don't do anything to deserve that treatment."

"Are you a shameless shill with the social skills of a sociopath?" This slipped out – Martin was a bit shocked at their reaction. This was no impressionable undergrad. Blunt had fronted hostile Senate enquiries.

Blunt and Wilson burst out laughing.

Wilson said, "That broke the ice. Sorry, I couldn't resist."

Martin looked thoughtful. "I don't know if I should really be showing people this. I don't want anyone to think I am approaching this with some sort of bias. Do you mind not telling anyone else about it? And I think we should delete it from your disk." He turned to Blunt. Blunt reluctantly dragged the file to the trash, and, as Martin watched, emptied the trash.

Blunt looked up. "There. And I won't even tell Ian."

Wilson nodded solemnly, and drew a finger across his lips. "Right, so lets talk about ice, should we?"

Blunt smiled. "Our favourite subject."

"Yes," said Martin. "I'm interested in why the IPCC doesn't model major sea level rises, and how big the threat really is."

Blunt nodded. "We went through some of this yesterday.

The biggest issue really is that the IPCC has not accepted any model of non-linearity."

"Non-linearity? "

"Temperature goes up n percent, melting increases more than n percent. Like compound interest."

"I know what non-linearity means – it's just a bit of a shock that this is what they are leaving out. I mean, if this is growing at anything significantly above linear – quadratic let alone exponential – won't we see a rapid acceleration at some point?"

"Exactly." Wilson nodded as Blunt spoke. "We can't model it accurately as yet because the system is so complex, but there is evidence in the paleoclimatic record as I mentioned yesterday of very rapid melts resulting in multi-metre sea level rises in less than a century."

"Why would this happen so fast? It takes a lot of energy to melt ice."

"Aha. My hobby horse. The West Antarctic Ice Sheet is inherently unstable. Half of it is below sea level, and not just a little – up to two kilometres below sea level. That's more than a mile."

"We do kilometres in Australia."

"Oh, sorry. I tend to think the entire English-speaking world is as backward as us ... sorry, Will. I know you're cool. Anyways ... there's a chunk of ice at the edge of the WAIS that's over water. We call that a shelf."

"Like Lars B?"

"Larsen B, you mean. Exactly. Will, your guys were there ..."

"Yes, we had a research vessel in the vicinity.

"Over three thousand square kilometres of ice broke apart in thirty-five days. Spectacular. We didn't have good models of this sort of fragmentation before, but the science is improving.

"Interesting. What is the mechanism?" Martin asked.

Wilson explained. "Melt water fracturing is the best theory so far. A little water pools on the ice. Ice reflects heat much more efficiently than water, so the water forms a warm patch which can go quite deep, and the water seeps into cracks. If it refreezes, it expands and cracks the ice ... the same effect as you see when water gets into a crack in the road, then freezes and cracks up the road. Since these are not common events, we have to work up a variety of models, fit them to satellite pictures and so on, then wait for another one, to see which model holds up."

"I see. And why is this so important?" Martin asked.

"I'll field this one," Blunt butted in. "The ice shelves are over water, but butt up against the land-based ice. If they break away, the land-based ice can also break away."

"And if the land-based ice is below sea-level ... ?"

"Then it could be subject to similar processes – and break up a lot quicker than we expect. But of course the dynamics are considerably more complex as long as the ice is in contact with the ground – there's friction, it's not floating, if part cracks off and starts to float, what's left behind will behave in ways that are even harder to model."

"I see. So this will take enormous computing power – you will have to model several variants to get a range of answers."

Wilson was intrigued. "Geoff, I thought you said he was a journalist. Martin, my boy, you aren't doing this as a sideline, by any chance, from an academic career?"

Martin smiled. "Thanks. No. This is all I am, for now at least."

Wilson nodded. "Pity. We are planning another study of the ice as temperatures warm in early summer. I'm sure you'd find it interesting. But it would be a lot easier to add you in as an academic. I don't think we have a budget to take journalists."

"Well, if I can swing something ... even a temporary researcher job at a university ... ?"

"Make it easier for me and I'll do what I can. But I think you should work on a plan to interview remotely and use stills and stock footage. The logistics of travel to Antarctic bases are nontrivial. Keep in touch ... my word, look at the time. Geoff: give him my Skype name and email would you, there's a good chap. Sorry. I must rush."

Martin waved as Wilson's image disappeared. "We've used up a lot of your time – if you can find a nice spot for the interview, let's go to it without wasting more time."

They did the interview outside the diner on the ground floor for local colour, Martin oblivious to hints from Blunt that he should know something about it, as he set up his camera.

"I am talking to Doctor Geoff Blunt, a senior scientist at NASA's Goddard Institute for Space Studies in New York.

"Dr Blunt, you have been outspoken on the dangers of climate change. What does the future hold if we continue to burn fossil fuels as if there's no tomorrow?"

"No tomorrow? That could turn out to be prophetic."

"Some have called you an alarmist. Could you explain

why?"

"I don't know what goes on inside someone else's head. If the science points to an alarming conclusion, must I keep quiet because it may be 'alarmist'? Would it be better to keep quiet and let matters take their course when you know better?"

"The specific thing you've written about frequently is ice melts. In their latest figures, the IPCC has reduced their estimate of sea level rises by the turn of the century. Could you explain where the problem is?"

"Yes, I can.

"The IPCC has *not* revised their sea level rise estimates down, contrary to the spin by the contrarian camp. What they have done is they've separated out the aspect with the greatest uncertainty, ice melts from the major ice caps. The reason for this uncertainty is emphatically not cause for complacency. These ice caps are melting faster than the models the IPCC supports, and the rate of melting is accelerating."

"But surely there is an enormous amount of ice there – even if it is melting faster than predicted, won't it take a long time to melt completely?"

"You would think so, but our biggest worry is the West Antarctic Ice Sheet. Half of it is below sea level, and not only a little below sea level – up to two kilometres. For those who do miles, that's over a mile below sea level. This means that melting at the edges could destabilize the whole thing."

There followed an exchange much as that between Martin, Blunt and Wilson.

Finally, Martin asked: "Dr Blunt: what would it take to develop a detailed model of the ice caps to make this all convincing to the sceptics?"

"Some people will not believe even if they see tropical rain forests in Alaska. I've not only looked at the instrumental data and worked on detailed models, but been to the Antarctic regularly. For me, there is no question. Based on what we know of the distant past and the fact that rapid change has happened before, multi-metre sea level rise by the turn of the century is far more likely than the IPCC is willing to admit. The contrarians claim the IPCC is alarmist. I say that their attempt at being 'prudent' is supporting dangerous complacency."

As they walked back inside, Geoff consulted his watch. "Well timed – I'll take you straight to Ian. And I promise: not a word about the Schoor interview."

Ian Grant shifted uncomfortably in his chair, as Geoff ushered Martin into his office.

"Ian, here's Martin Truscott. Sorry, must rush. I think you two might get on OK."

Grant looked at Martin. "No camera?"

Martin patted his backpack. "I'm not planning an interview – more interested really in checking some background, if you don't mind – there's just so much I can pick up from reading."

"OK. Let's get some lunch then. You watch Seinfeld?"

"Not really – I don't watch much commercial TV."

They went down to the diner.

As they went in, Grant pointed at the sign. "Tom's Restaurant. This is as far as it continues to look like something from Seinfeld. They did their own interior. And changed the name on the outside."

Seeing Martin's indifference, he added, "Anyway it's convenient, fast, inexpensive. Let's get in quickly before the student crowd."

Martin ordered a bagel with lox and cream cheese, never having tried the combo, while Grant ordered a burger. The orders arrived fast, Martin's with a generous filling of smoked salmon.

Over food, Martin checked some details in his notes, then: "I've been trying to understand this issue some people have with accuracy of weather stations – the photography bunch. You know, the claim that global warming is really a data error, failure to take into account the urban heat island effect."

"Yeah, like photographing a weather station proves anything. Its placement may look bad, but you need to do a thorough calibration, and compare it against its historical record. Remember, the thing we are looking for is a trend. That's why we measure 'temperature anomaly' versus a specific period, not actual temperature. If it's been like that for twenty years, and nothing has changed the trend, it doesn't matter if it's inaccurate – as long as its consistently inaccurate."

"Wouldn't it be better though if problems like air cons venting onto weather stations were eliminated?"

"Well, think it through. In the US, we have 1,221 weather stations. Now, these guys photographing stations are finding the ones in relatively convenient locations. To do what they want, we'd have to pretty much check them all. To do a thorough calibration is a good day's work. Add in travel time and costs, you average maybe 2 a week – take you over ten years to do the whole job. So you put in 10 teams instead and get it down to 1 year. Even if each team is only one person, you are talking a cost in the millions. Meanwhile, you get to the end of the job, and someone else puts a barbecue in next to a weather station that was OK before, and you have to start again, because you missed that one."

"So can't you isolate out the most obvious cases?"

"We do. We have sophisticated data analysis techniques. If a site is varying from the behavior of its neighbours, we reduce its weighting in the models. And saying that we ignore the urban heat island effect is rubbish – it was in the very earliest attempt at modelling global warming by Guy Callendar in 1938. Oh, and if we suspect that urban areas tend to be warmer than average, how do we measure that unless we have weather stations in urban settings?"

Grant held up a finger, hesitated a second, then said, "How do they conduct a census in your country?"

Martin tried to remember the details. "I think someone drops off a form. You can either fill it in, or do it on a web site. I'm not sure how they collect the form – I did the web version."

"And who controls the quality of the data capture?"

"I guess ... pretty much the person filling in the form."

"So let's say there's a neighborhood – let's make up a name – Crackton – that's fallen from yuppie hangout to drug dealers shooting it out. Illegal immigrants move in hoping the drug dealers will keep the cops out, and maybe half the original population is stuck there, living in fear but unable to move. What will the next census show?"

"I dunno, maybe a fifty percent drop in reported population."

"Exactly. So what do the census people do? Allow the error to slip through? Check for anomalies, like changes beyond the norm, using mathematical techniques, as compared with the general trend?"

"The latter, I guess."

"Then what? Do they send in the police to count everyone?"

"Probably not – the police may be paid off by the drug dealers."

"Exactly. And if the census becomes associated with a police raid, guess how accurate the data will be next time around. So they use a variety of data analysis techniques to compare with other similar areas, and correct. You see, Martin, one of the biggest advances in science in the twentieth century is figuring out how to work with noisy data – cleaning up by statistical techniques."

As he paused to catch his breath, Grant noted a change of pace in the restaurant. "I think the waiters are getting edgy, the place is filling up. Let me show you some pictures in the office that will erase all doubt."

As they walked back, Grant added: "One thing these guys can't answer. If the temperature trend is just bad data, why are sea levels rising?" At his office, Grant pulled up a couple of web pages, and put them side by side. "This one –" he indicated a world map coloured in shades of yellow, orange and red – "shows the global temperature anomaly averaged over last year, and this one –" indicating another map of the world in shades of blue – "is population densities."

Martin put up a hand to stop Grant from going on. "Don't tell me the colour coding – let me guess. The Urban Heat Island Effect is strongest in the Arctic, because that's the area of highest population density."

Grant laughed heartily. "This is the thing, you see. The so called sceptics claim that our data is all wrong, our models are all wrong. But the clear evidence when we measure the temperature anomalies is that the highest increases are nearest the poles – tell me how that's an error caused by an uncorrected urban heat island effect."

Martin nodded. "That's pretty conclusive – let me get those web sites."

"OK. And the thing you need to particularly note is that the Arctic has heated the fastest until recently. Now the Antarctic is catching up – if only in a small patch for now."

"Penguins with barbecues?"

*   *   *   *   *

JFK Airport. So this was it – a short hop back to Boston to make the connections home.

Martin found a space to set up his computer, and burnt a few backups to DVD. "Just hope these are good – no way am I showing them off in a public place."

He packed the DVDs away, evenly splitting duplicates between his backpack and the computer bag.

8 Home Again

H E TRUDGED up the stairs. There it was again: the front door he'd opened alone so often. Everything suddenly caught up with him: the intensive days in the US, the need to put all the stuff together – and yes, he had to go back to McCarthy to finish that one off. Nothing recorded. What if Schoor had put him off?

And Angie. Had that been real?

Now he had to contemplate the concept of talking to Julie after ... how long had it been? Four years? She had been so distant, so unwelcoming that time, he had hardly said hello before he wanted the conversation to end.

He checked his watch. Just past 9 pm. What time was that in California? Boston was something like 14 hours behind Brisbane. How much more was California? His brain hurt. He gave up on dizzying computations, found his keys and opened the door. He put his backpack down and turned on a light. The place was as he remembered it except for a patina of dust.

"You'll help me, won't you." Computers aren't people, but his Mac had been his constant companion for the trip. He fired it up, and switched the time zone. He hit LA on the map. "Close enough." It showed 4 am plus a few minutes. He clicked on the east coast. 7 am. "I wonder if she's still in Boston time? Naah. Better not wake her up."

He started to unpack. He was tired from the trip, but it didn't feel like bed time. A cheap ticket doesn't give you the most direct flight. The Qantas flight was passable, but getting as far as LA was a real killer. "How the hell can they have a power failure that puts air traffic control out at an airport that size?"

Getting as far as LA on US Airways had seemed a fair option when he booked the ticket, but a no-frills airline is not well equipped to make an unintended 2-hour layover at a tiny airport bearable. Then, the nightmare security at LA, and missing his flight to Brisbane . . . Luckily his gear all just made it into carry-on, otherwise he'd still be at the airport trying to find his luggage.

"What to do?" He looked around the apartment. There was his basket of dirty laundry, to which he could add some from his backpack. He put it all together, and went down to the laundromat downstairs.

Watching the clothes in the drier was strangely relaxing, something he hadn't noticed before. Perhaps he was starting to get his life in order if something so simple could feel so harmonizing. Red sock. Green shirt. White sock. Blue jeans.

By the time he had lugged the basket back upstairs, he was really tired, and collapsed on his bed, still in his clothes.

He woke up hours later. It was still dark. He tried to sleep again, and eventually gave up as the sky started to lighten, characteristically early – Brisbane is too far east for its time zone, and it gets light early even when it's not summer. 5am. About midday in California. He opened his computer, checked for mail.

There it was: something from Angie.

```
I hope you're back home -- I tried to get
you on Skype, but your computer was off.

Let me know what's happening.

Miss you.
```

He opened Skype. There it was, the missed call. When he should have been home, if he hadn't missed the flight. "Dammit. I should have checked mail last night. What was I thinking?

"Nothing, obviously."

He clicked on her name.

She was on line.

He started a call.

She picked it up almost instantly. Her face appeared, pixellating as the bandwidth varied from adequate to unable to keep up. He silently cursed his inability to afford a faster broadband service.

Meanwhile she was talking. "Hi, where've you been? I tried to reach you as soon as I thought you'd be home."

"Stupid LA airport. The air traffic control was down, so I was diverted to some tiny air strip – Grand Rapids Colorado, or something."

"Grand Rapids is in Michigan . . ."

"Well, Grand Junction, or something like that. Anyway we were stuck on the ground for two hours, and I missed my flight. Only got home after nine, and I guessed you would be in bed at 4 am."

"I admit, I gave up waiting about 2 am."

A pause. He wasn't used to someone staying up for him.

"I'm sorry. I really, really wanted to talk to you, I didn't even think of opening up Skype, or even email. Still a bit tired."

"I bet. You look a mess."

"That's just my low bandwidth connection."

They laughed. "No, it's not. Did you talk to Julie yet?"

"Nope – but I will today. It's a work day, so I'll try to catch her before she goes out. I'm not sure where she works, unless she didn't change jobs."

"What time is it there now?"

"5 am, 3 pm east coast time. Couldn't sleep, even though I was tired."

"What time are you going to call her?"

"I was thinking something like 7:30 – not so early that I'd wake her up, not so late that she'd be on the way to work."

"I'll stay near the computer in case you need moral support. In the meantime, get yourself a nice espresso, take a shower, whatever makes you feel better. And shave off some of those rough pixels." There was a forced cheerfulness in her voice – or was it just Skype breaking up?

He laughed. "And you, how are you doing back home?"

"OK – signing up for classes, catching up a bit on sleep, doing a bit of shopping. It feels weird preparing for classes without having to worry about the grades. But old habits . . . I guess I'll get back into it. Went for a swim this morning – only a few seconds off my regular time. Feels so strange to be on my own again. I can't believe we were only together a few days."

"I did dump my whole life on you."

"Not really, only Julie. You must tell me about family some time," she said. *Was* there a tension in her voice?

A long pause. "You're right. I hardly know anything about you too, except I like being with you."

"Yeah, I miss you too. Take a shower, get yourself in shape, talk to Julie then let me know how it goes." Nope, sounded totally in control. Must be imagining it. Why would she be tense anyway?

"Bye." He broke the connection slowly.

He rummaged through the sparse contents of the freezer, and found a bag of coffee beans. He took them over to the grinder, ground a good scoop of them, and made an espresso. "Do what the doctor ordered."

He felt a bit better after that.

"Man. She's good. Better do the other thing."

The hot water felt like needles, but he left the shower feeling ten years younger and a good bit more awake.

"Still time to kill. I wonder if I can find a newspaper."

He went outside, and found a convenience store – not surprisingly, where it had been two weeks before. He bought a copy of *The Australian* and, unusually for him, because he despised its trashy tabloid style, the local *Courier Mail*.

The papers didn't have much interesting news, but both had sudoku games, and *Courier Mail* had a few other puzzles, which he battled to tackle without falling asleep.

Eventually, he gave up and made another espresso.

The clock meanwhile was slowly ticking on. He looked

up Julie's number in his old address book, and wondered if she was still in the same place. "No reason not, I suppose. It was a nice enough place." He looked her up in the electronic white pages. No listing. So ... try her old number.

The second hand on his watch clicked past twelve, and he slowly punched the numbers.

Amazingly, the familiar voice answered.

"Hi," he said, at a loss, but she recognized the voice from the one syllable.

"Oh, it's you."

"Sorry to call you so early, but I was hoping to catch you before work ... are you still doing that advertising stuff?"

"Well you did catch me before work." There was a hint of irritation in her voice, a kind of distance he had not sensed before, even when she was blowing him away.

"The thing is, I need a favour. I'm doing this pilot of a TV show on climate change – how the science stands up to sceptics. I thought the sceptics might have a case, but I am pretty sure now that they don't. I have some killer material but I don't have a high-end editing rig and it will take me weeks to do this on mine. So I was just wondering ... do you have stuff at work I could use? I could do it over the weekend, not get in anyone's way ..."

"Martin, this is not so easy for me to do. I have a family." This was an unexpected development, something he had never thought about; this put a new finality on their separation. "I can't just drop everything and spend a weekend at the office."

"Oh. OK. I'll find a way –"

"I didn't say I wouldn't do it. Just give me time to try to

work it out, OK?"

"Thanks, I would really appreciate that."

He wasn't sure who ended the call.

All the old doubts came flooding back. This was so much like times in the past when she was going to call him back. He sat for a while, knees pulled up to his chest, a near-foetal position. "No! Dammit, she isn't going to get to me like that."

He went to the computer. As promised, Angie was still on line. He started a call, and she picked up almost instantly again. He started to feel better as her face resolved into smaller pixels.

"You called her," she said. "Didn't go well?"

"No. She told me she had a family and couldn't just drop everything."

"Well, that can't be totally unexpected. I mean, how many years has it been since you've seen her? She could have a three-year-old and you wouldn't know. So she said no?" The Skype connection was better – whatever the problem was before had gone away.

"Not exactly. She said she'd call me back. But she used to do that a lot."

"And this felt the same?"

"I don't know. I really don't know. There was a distance in her voice that I never had from her before. But it's been years since we've even talked. I was surprised that her old number still worked – especially with the family thing."

"Maybe she moved and kept the number."

"Maybe ..."

The phone rang.

"I'd better get that."

It was Julie again.

"Listen, I had a chat with James. We both really admire the way you've been taking on the big corporates in your work. He's agreed to look after Donnie the whole weekend if that's what it takes. I'll have to clear this at work, but they'll do what I want."

James was a name he didn't recognize. "Does James know about ... *us* ... ?"

A long pause.

"Yes. I told him you used to be a really important part of my life, and if this was the last thing I could do for you, he'd better support me."

"Wow. I don't know what to say."

"Saturday, around nine? Do you still live in the Valley?"

"Yes."

"Should I pick you up? The office is in the city. River-side."

"No, I think I'll walk. It will clear my head."

She gave him the address.

"Still the same place?" He asked.

"Yes, it worked out surprisingly well. I never pictured myself in advertising."

"And I never pictured myself making TV documentaries," Martin responded.

"See you Saturday then," were her final words.

Someone ended the call. As before, it was all a bit hazy. Then he heard Angie in the background.

"I didn't hear much but that sounded sort of OK to me."

He moved back to the computer.

"Yes, yes. I was a bit surprised. I really thought she didn't want to know about me anymore."

"Do you think you'll be OK dealing with her?"

"I *think* so. I hope we can keep talking. You're my lifeline."

"Of course. I told you already I miss you."

"You did too. Twice, in fact. Didn't I tell you the same thing yet?"

"That I miss me?"

They laughed.

"Meanwhile," he added, "I have other business to attend to. I need to get back to McCarthy to record something, otherwise the stuff we'll be editing over the weekend will be a bit thin. I've just been worrying that Schoor has put him off."

"You don't have much time before the weekend – talk to him soon. If Schoor can influence him like that, that's only to his discredit."

"Maybe so, but – fuck it, you're right. I forgot about the international date line. It's Friday already. I hope I can catch him today." He was looking at his watch in some agitation, the day of the week suddenly in focus.

She smiled – he imagined it was the usual impish one, if some of the detail was lost – "Better catch him before he goes to work."

"He's retired."

"I know that. Tell me how it goes. And lose the potty mouth. You don't want him to think you're a boorish dolt."

"A boorish ... sorry, Ange, I don't usually talk like that." *When anyone is listening*, he added to himself. As he broke the connection, she was peering at him closely as if to see if the pink tinge to his face was embarrassment or a glitch in Skype.

It was now close to eight, so he decided to call McCarthy straight away. The phone was picked up on the fourth ring, so at least he wasn't a late sleeper.

"Hi, Martin Truscott – do you remember my documentary project?"

"Yes, of course. You seemed a pleasant enough chap, but Schoor tells me you were pretty dreadful."

"There's two sides to this. He was very rude to me, and refused to answer questions on the detail of the science. This thing for me has been about the science from the start. Is the science good? Does it stand up to detailed scrutiny? That's all I am after. I thought we had a very good discussion when I visited you."

"I thought so too. If you are fair dinkum and this is not some kind of ambush, I am happy to see you again. Everyone knows my views anyway. There's no way you can get away with selling me short."

Martin reassured him. "I have no plan to do that. Look: I will let you see my final edit and if you disagree with anything, I'll either edit it to your satisfaction or give you space to rebut. Nothing from you will be in the final edit, unless you are happy. Can you live with that?"

"Sounds more than fair. When did you want to do this?"

"I've managed to line up an editing session over the weekend, so if you have time today, that would be great. Sorry about the short notice but my travel has been hectic."

"Short notice? Well, I don't have that much to reschedule, I suppose. Can you get here by 2pm? I have some friends over for lunch, and this would be a good way of breaking the party before it gets too tedious."

"Great, thanks. I appreciate your help, as always."

"Yes. But I'm not sure if I can help you get access to any more sceptics now that Schoor has put the word out."

Martin organized himself a rental car, parked it outside, then went back to the apartment to work out what to take.

He woke up on his bed, sun streaming through the window. "How did I land up here?"

His second action was to check the time. Then, on his computer, the date.

"Fuckit! At least it's still the right day." He shook his head to clear it, remembering the awkward moment with Ange. "Better not get into the habit of cursing at people. The benefits of living alone ... "

He ran to the bathroom contemplating the benefits of *not* being alone, did a quick shave, then ran around the apartment rounding up things he needed to do the interview: camera, computer, printed copies of papers, notes ... yes – the notes from last time were there.

Coffee to stay awake? "No time." He grabbed a couple of whole coffee beans and chewed them, the caffeine hit shocking him into greater wakefulness.

9   McCarthy

THE TRIP NORTH was much as Martin remembered it, which was just as well because he was battling to stay awake. His panic subsided as he remembered he'd allowed an extra half hour because of the unpredictable traffic. Not only was the Sunshine Coast a popular tourist destination – possibly less so late spring than in high summer – but many Brisbanites chose to live out there, despite the commute from hell.

Luckily, the timing was on his side: not that many commuters headed out there around midday, and even for Brisbane, 12:30 Friday was a little too early to quit work for the weekend.

The sky was an almost unnaturally bright blue; the grass unusually green for that time of year, the driest part of the dry season.

Tired or no, it felt good to be alive. He had Angie to make him feel good about himself, and Julie's reaction had been unexpected – if balanced by the shock of the words, "I have a family." No, he had to put that behind him. If only . . . he just didn't understand. Her words were almost admiring, yet she'd shoved him away so hard, he still felt bruised. The picture of her in his mind was hard to put aside, so he focused on Angie – on some of the ideas they'd talked about for finishing the movie.

That got him in a new train of thought: what would he talk to McCarthy about? He'd been in such a rush, he hadn't even sketched out a list of questions. All he had was the papers he'd been reading all along, and his notes from last time – and what felt like a lifetime of experience. No ... he knew what to talk about: same as before, but with a deeper probe. He checked the time. "Making good progress. If I get there in good time, I can write something down.

"But, meantime, let me think ... keep awake ...

The interview got off to a false start. McCarthy was noticeably nervous, compared with the first meeting. The bead of sweat on his forehead didn't have anything to do with the warmth of the back yard, Martin was willing to bet.

After a few questions, largely echoing their earlier conversation, Martin said, "It's a pity we didn't do the interview first time – it went smoothly then. How about we take a tea break, then try again?"

McCarthy visibly relaxed. "Good idea. I must say, you don't seem any worse than last time. Maybe Schoor did overreact a bit.

"What kind of tea do you like? Earl Grey, English Breakfast, something herbal?"

They walked back into the house.

"English Breakfast sounds good – no milk or sugar, thanks."

Martin sat looking at the books in McCarthy's living

room. Most were about paleobiology, but he also had a
fair collection of classical literature, detective stories and art
books. There were a few pictures on the walls – local scenes,
as far as he could tell. He was examining one as McCarthy
returned.

"Don't pay too much attention to those – just a pastime of
mine."

"Oh. I was going to ask who the artist is . . . "

McCarthy made a gesture of modesty and put the tea tray
down. "How strong do you like yours?" He started to pour.

"That looks good, thanks." Martin took his cup, and
turned to McCarthy as he started to speak.

"So what happened with Schoor? He told me you really
went after him. You look about as tame as I remember you
last time." McCarthy looked Martin up and down as if for
evidence of a stunt double.

"I had some doubts about his science. He didn't like being
questioned on the detail."

"I see. But why was your interview so aggressive?"

"Schoor set the tone. He was abrupt with me, and treated
the whole thing as if it was a waste of time."

They drank their tea in silence, then McCarthy looked
Martin straight in the eye. "Should we continue then? It
seems we are covering pretty much the same ground as your
first visit."

"That's right. I have a few follow-up questions based on
talking to the others, but nothing that should present a big
challenge."

Tea and reassurances out of the way, they went outside
again. Martin covered previous ground: McCarthy's aca-

demic history, how McCarthy joined the debate, whether 1998 really was the end of global warming ... He pushed McCarthy a bit harder than on the first visit on the urban heat island issue, but McCarthy wouldn't budge. He insisted the data was unreliable.

All really predictable stuff ... so Martin tried to find a new angle.

"Professor McCarthy, the thing that's troubling me is that when I really get into the detail, the sceptic position has a lot more gaps than the mainstream position. Schoor's approach appears to me to be mostly statistics, without a physical model underlying it. Hayes only really claims to have found greater uncertainties than the IPCC admits to.

"If the situation is less certain than the IPCC's models, isn't that cause for more concern, not less?"

"Did I ever say I wasn't concerned? The big issue for me is that we do not destroy our economy over an issue with so much uncertainty."

"But isn't it correct that uncertainty on the pessimistic side is also a concern? That the IPCC models could be excessively optimistic on the scale of climate change?"

"I don't think so – I doubt it very much."

"Thank you professor."

Martin turned off the camera.

They shook hands.

"Professor, that wraps it up for my pilot. I hope I can get the full project funded, but I do really appreciate what you've done for me. I hope I didn't wreck a good friendship with Schoor."

McCarthy rolled his eyes. "I can't say he's a friend as

such. But, you know, robust debate is not something that's ever frightened me – even if you thought it did when we started. Though I do appreciate your willingness to let me collect my thoughts. I'm surprised he took it that hard."

He walked Martin to the car.

10 Edit

MARTIN FOUND THE BUILDING easily, with fifteen minutes to spare. With a bit of time in hand, he wandered down to the City Cat stop to watch the river for a few minutes. The Brisbane river is a tidal river – quite far upstream – and he never tired of watching its various moods, a thing he suddenly realized he'd learnt from Julie.

A City Cat arrived – a big blue catamaran-style ferry, with glassed-in cabin and open-air seating on the fore and aft decks. The arrival of the boat made him conscious of the time. "Better get moving."

He walked back to the street, and found the entrance to the building. She was there a few minutes late, panting from exertion. "Damn parking. I forgot the building was closed on weekends, and city parking is such a chore."

She stopped close to him. "How *are* you? It's been so long."

"It has, hasn't it. You still look great."

"And you still look as if you should still be in high school."

She found a key and opened the glass doors. "I don't

usually visit this place on a weekend, so let's hope I have the alarm instructions straight."

A few minutes later, they were in the company offices, an expanse of glass facing the river.

"Nice. I bet the rent's not cheap."

"No. But neither are our rates."

She showed him to an expansive desk with a good view of the river. "This is where I sit when I'm not creating." She smiled wryly. She pointed to the centre of the room. "We'll be working there. The lighting's more controlled away from the windows." They walked over to a wall where she hit a row of switches, turning the lighting up, then went to the centre of the room, where a couple of large Mac Pros with a pair of gigantic screens apiece sat humming gently.

She hit the keyboards on both, and typed in a password on each. "We'll work as me – no point in creating a login for you just for the weekend – we can delete all traces afterwards if you want to keep this confidential."

"Well, I wouldn't like it to get out until I have a bit more in the can. I've got lots of DVDs to back up on.

"One more thing before we start. How secure is your setup here? Word has already gone out that I gave one of the interviewees a hard time. I don't want too much to leak out, otherwise it could spook others away from being interviewed."

"You have no idea how competitive advertising is. We have to work with clients who could lose millions if something leaks before they are ready to hit the market. We have encryption on the hard drives, military-grade secure deletion of files, firewalls between everything let alone the outside world. If we just make sure we delete everything when we're done, there will be no problem.

"And one more thing – have you got a logo to watermark the copy you give to Prentice?"

"Why?"

"These things can so easily slip out of the system. Next thing someone's using your material without permission, and you have a hell of a job fighting copyright lawsuits, unless you're a big corporate. Once you've sold it you can give them a clean copy."

He thought for a bit. "There's a kind of bird I used to like watching in the garden. I'll look for a picture tonight."

She looked a bit surprised, as if this was an unexpected aspect to him. He smiled, and she asked, "So are you happy now that it's all going to be secure enough?"

He nodded, and sat at one of the machines. "What are we using here?" She leaned over and used the mouse to launch something.

His eyes widened when he saw the range of tools. "Wow. This has a hundred times the features of iMovie."

"Well, duh. It also costs 4 squillion dollars more."

"I know. I just never had the funds to pay for something like this. Not even to rent."

"Let's start with your clips. Give me some to look at, and do some yourself, and we can swap ideas. Do you have a plan for this?"

"Yes. I have a sequence here –" he pulled a sheet of paper out of his backpack – "and I wonder if you could shoot a few linking scenes of me talking."

"Of course – the river backdrop will look good. It'll just take effort to make the lighting work."

"Great. But I'd like you to see some of the material before you look at the plan." He handed her a couple of DVDs, and took a couple himself. He put the sheet of paper down where she couldn't see it. They worked in silence for a few minutes.

The first one he loaded was the interview with Schoor. He said nothing, trimming a few unwanted details, then hit *Preview*. She looked up as the sound started, then she was behind him, watching in silence.

As the camera followed a trembling, retreating Schoor, she said, "Wow. Who was that?"

"Schoor. Harvard professor. Up to his gills in fossil fuel money. Publishes junk science and uses Harvard's name to stoke up the denial camp."

"Ah – that would be the one someone thinks you've been mean to – or are all of them like that?"

"Just the one. I admit –" he displayed the impish grin he'd learnt from Angie – "it was a bit of an ambush. I had an introduction from McCarthy, our local sceptic-in-chief, so probably he was expecting me to be a mug. I didn't like his work. I didn't like his funding and I didn't like his attitude. Luckily for me, he played the part perfectly."

"Luck? You set him up."

"Yes, but this guy has testified many times before the US Senate. He should be able to handle the pressure. It's not even as if I am a pro at this game."

"Aren't you?"

"You are. You tell me."

"I've only seen two clips, and what I've seen so far is

pretty good. Science docos are much, much harder to make than ads. You have to communicate something much deeper than a sound bite."

"Is that all you do these days: sound bites?" He sounded a bit disappointed.

She looked down. "I'm afraid so. I feel such a sell-out."

He put a hand on her shoulder. "Come on. You aren't selling drugs to kids or anything like that, are you? This is just a job."

She still looked at her feet. "When we were students, I was the idealist. Nothing was good enough. You were the quintessential nerd, only interested in geeky things. Computer games, programming, faster graphics cards."

"So why were you interested in me? I mean you picked me up at that party, the thing after that bunny hugger talk. I did nothing: I've never picked up a girl in my life."

"I had to rescue you. That pale skin, dark rings around your eyes: you looked like a raccoon."

"I did *not* ... well, maybe a bit ... "

They both laughed.

"So I was some sort of project then, and you dumped me when it was over?"

"Oh Marty, no, no. It may have been a bit like that at first, but I quickly found the real person under the geeky shell. I've had lots of boyfriends before, but no one quite like you. You were the best."

He looked quizzical. "But why ... ?"

"When I got this job I felt such a fraud. I'd dragged you out of your comfort zone. Made you abandon all your hopes and dreams for a life of principle, then I landed this cushy

job that contradicted everything I stood for. And I had ruined
your big life plan. I mean, who was I to pull you away from
computer gaming, even if I thought it was stupid? It was your
life. And there you were, left without any marketable skills,
and there I was, doing something I despised, because I wanted
financial independence."

"Julie, everyone gets a job. Well OK, those with some
sort of useful skills ... it's kind of normal for students to be
a bit out on left field, but to move back to reality when they
graduate. I didn't expect you to be a member of the leader-
ship collective of bunny huggers international or whatever for
life."

"At the time, I didn't think you would see it that way.
Every time I saw you, I saw someone whose life I had trashed,
and I had no answers."

There was a silence. He realized his hand was still on
her shoulder, and slowly withdrew it, feeling awkward. Then
something occurred to him. "Maybe Angie was right," he
said, almost whispering.

"Angie? Do tell."

He looked even more awkward.

"It's OK. Tell me. You don't mean to say you were all
lonely and miserable these past five years? ... What was that
about 'never picked up a girl in your life'? You haven't been
lonely and miserable haven't you?"

"Of course not," he said, a little too vehemently.

"I'm so sorry."

"It's OK. It's not your fault I'm so useless."

"Useless? Tell me about Angie."

"I met her in Boston at the hostel where I was staying for

this –" he gestured at the now still movie clip.

"Hostel?"

"Yeah. This is just a pilot, remember. Real low budget. If I can make it look good, I can get a real budget to make the full thing. That's why I so needed a way to edit the thing fast. Anyway, I was living as cheap as I could. So, Angie . . . I was stupidly reviewing that Schoor clip in the hostel TV room where anyone could have walked in and seen it, and, luckily for me, someone nice walked in and saw it."

"Have you got a picture?"

"Only a screenshot off Skype." He pulled out his computer and showed her.

"Wow. She looks like a babe."

"Well, yes, but she's also pretty bright, and great company. Very energetic, swims for recreation."

"Lucky boy. Don't let this one get away. Take up swimming so you can keep up."

"I will. All the way across the Pacific, if I have to."

There was a pause.

He looked her straight in the eyes, the old physical communication back at last. "We've got a movie to make."

*   *   *   *   *

Sunday, 4pm.

"I'd say that's pretty good for a pilot. I'd even run it as is if you can't get more material."

"Thanks, Jules. Thirty minutes is a bad length though – too long for a 30 minute show with ads, too short for an hour. It will have to go up or down, and I am missing too many of the real authorities to go down."

"Let's play it through one more time to see how it looks. The watermarked copy, since we reviewed the other."

He nodded, and she started it running.

It started with white titles on a black backdrop. Very simple, minimal FX. In the bottom right corner, one thing didn't vary – a translucent image of a bird with a long tail held high.

No Tomorrow

a Martin Truscott Documentary

Clip of Geoff Blunt saying, in response to being asked about using fossil fuels as if there's no tomorrow: "No tomorrow? That could turn out to be prophetic." More credits.

Switch to Martin, filmed in Brisbane (right here, with Julie doing the camera): "My name is Martin Truscott. One thing I abhor is dishonesty. When Martin Durkin's movie, *The Great Global Warming Swindle*, screened in Australia, I was annoyed that it was so badly made. The scientific claims he used to make his case could so easily be taken apart.

"What if Durkin was right – that the science is flawed, but he did a poor job of making the case?

"This film is about exploring exactly how good the science is – on both sides.

"And for once, the focus really is on the science."

There followed a sequence of clips – Wilkinson, McCarthy, Schoor, Blunt, Hayes, then back to Blunt – with edits of stills and cuts back to Martin. A paper by Schoor, with funding acknowledgments highlighted. Wilkinson's 3D models. Scientific papers, with graphs prominent. Stills of the

Antarctic, showing ice reduction.

Finally, Martin was back on camera.

"There we have it. The best the self-proclaimed sceptics could offer. The sun accounts for everything. I couldn't find evidence to support this. The IPCC models leave out critical details. They are all there: everything the sceptics claim was left out. Water vapour and clouds account for everything. I couldn't find evidence to support this. The IPCC models are all statistical manipulations, without reference to the physical world. All the evidence I found pointed the other way – that the sceptic positions were hard to support once you start looking at the real world. Finally, the sceptics are trying to focus all our attention on the possibility – note, the *possibility* – that the IPCC models are on the pessimistic side.

"What they are all ignoring is the possibility – and I emphasize again, *possibility* – that the IPCC is in fact erring on the side of caution, caution in predicting dire consequences of uncertain probability. That we don't fully understand them doesn't mean that the more dire consequences can't happen.

"I am not making a judgment here – just presenting the evidence. But if we want to be sceptical about the IPPC's modeling, we have to allow the possibility that reality may be worse than we want it to be, not just better.

"There's the evidence. You be the judge."

She was standing behind him, and nodded slowly. "Good work. I think this will get you your funding."

He turned to her, smiled, then frowned. "One thing I've been wondering about ..."

"Yes?"

"Why *this* job? I mean, it's true they made you an offer

but it was about the first interview you did. Couldn't you have strung them along a bit, and checked out a few more?"

"Marty, you never met my parents –"

"Well, they did live out in Bundaberg, and when you went to visit them in summer, you didn't exactly invite me ..." There was an awkard silence.

"You never talked about your family at all."

"True. So what happened? Did they force you to take the job? Gun to your head?"

"No. My parents were really struggling financially. Dad's business failed, and they nearly lost the house. Through all that, they insisted that I shouldn't work while studying. They told me uni was a full-time job. When I graduated, I just couldn't make them carry me for a minute longer than I had to. I was offered this job with a decent starting salary. I don't know how – I suppose if you are in the communication game, you have to come across well, and maybe I did that in the interview. Whatever. So here I am.

"So what about your family?"

"My parents split when I was too young to know. My mum was this distant figure, very affectionate the little time we had together, but my dad didn't want to know her. I tried to find out what had happened between them, and got a different story every time. I guess this is where I developed my hatred for lying. I just can't take it."

"Did you think I ever lied to you?"

"I don't know. Well, yes. When things were falling apart ... but, you know, I'd rather leave that, not go back there. That's all past. You can't go back."

"So you think your dad was lying to you. What about

your mum?"

"She died when I was fifteen. I never got to ask her."

"My poor Marty. You never told me."

There was a silence. Then he grinned. "Look at the two of us – both communication professionals, and we couldn't even talk to each other."

She relaxed a touch. "What about your dad? What about some communication there? Maybe he had reasons for the way he behaved. Did you ever ask him? If you can forgive me ... "

"No. No. But I think you're right. This is my week for getting my past straight."

She looked at her watch. "James and Donnie should be here any minute. Would you like to meet them?"

"Yes." He hesitated. "But maybe not yet. I need a bit more time."

"At least we can be friends again, can't we?"

He put both hands on her shoulders and nodded slowly.

She gave him a kiss that felt a bit more than "friend". She took a hand off her shoulder, and held it for a few seconds. "Better go. Let's pack up and delete everything."

"Making sure we have backups." He checked what was on his backup DVDs.

Deleting the files was the one thing that took a lot longer than he expected. Normally, deleting a file just means moving a little information from one place on the disk to another, making it disappear from its usual location and reappear in the trash. Military-grade deletion involves not only this but also writing random information several different ways over the area of the disk occupied by the files, so not even an expert with hardware probes could work out what was there.

While this was going on, they packed all their stuff. When he was satisfied that there was no trace of his materials on the disks, she dragged him out, and set the alarm.

As they reached the exit, a car arrived. She turned and waved as she opened the door. He waved back, then looked the other way.

11   Sale

I T SEEMED LIKE YEARS since Martin had been in Bill Prentice's office. He looked around – the industry para- phernalia he'd failed to pay attention to the last time mutely telling a tale of vast experience.

"So, you're back." Prentice caught his attention.

"Yes. Here's the pilot." Martin handed over a DVD.

"Yeah, well I'm busy now. I'll have to review it tonight. Not promising anything."

"Right." Martin nodded tiredly.

Martin started to back out of the office, disappointed that he would have to wait. Then he paused. "Oh, and by the way, I should really thank you for backing me. If it had been my money, I would probably have said no."

Prentice's face softened for the first time. He gestured around the office. "Look at all this, kid. This didn't come from nowhere. I had to start too. I have a good memory. I know what it's like. I gave you a shot. If it's good... that's the only way you get your foot in any other door.

"Here." He pulled out five dollars. "Have a drink on me. You look like you could use some R&R."

As Martin turned to leave, Prentice added, "I'll call you."

"Thanks." Martin felt weak at the knees. This was it. He was on the bones of his backside; only about one more grocery shop, one month's rent left in the bank. If Prentice didn't like it, he didn't know who else would. Maybe pitch it direct to the ABC, see if they would underwrite it? Worse: what if Prentice didn't get back to him soon? Should he wait a week and call him?

But Prentice was right. He could do with something to help him unwind. Instead of going to a bar though, he used the money to buy a couple of beers at a store, and stopped off for some groceries, and took them all home.

Then he sat contemplating the dust, the debris from his trip, the pile of DVDs.

There was some missing business to take care of.

He picked up the phone. Was dad still working at the same place? He thought for a moment, and started pushing buttons.

"Truscott and Partners Civils, how may I direct your call." Wow, maybe dad has bought the business?

"Is George Truscott available?"

"I'll check. Who may I say is calling?"

"Martin. Martin Truscott."

There was a brief silence.

"Marty! I thought you'd been abducted by space aliens."

"No, dad. I just thought it would be nice if we could get together, maybe dinner at my place tonight?"

There was a long silence.

"I've got artichokes."

"Son, you don't need to entice me. It's just a bit of a surprise. It's been so long."

"I know. A lot has happened. Can you make it? Around seven?"

"Of course. Do you still live in the same place?"

"Yes."

After the brief goodbyes, Martin looked around the place again, this time with no excuse not to tidy up, except one: he had instructions from Prentice. The beers were still cold. He cracked one open and drank it contemplatively. He put the empty in the sink, then started to hunt for cleaning materials.

Two hours later, the place was sparkly clean, the pile of DVDs neatly stacked, the computer winking serenely in sleep mode in its corner.

He took the second beer, colder for having been in the fridge for two hours. It tasted especially good. Then he realized he hadn't eaten all day, and the beer had hit him harder than usual. He collapsed on the couch, and woke up to the doorbell.

It was dark.

"Oh crap. That will be dad, and I didn't start cooking."

He pulled himself out of the couch, joints creaking, and hobbled to the door. "Man, I don't know how I got so stiff."

He opened the door, and there was the old man, silhouetted in the street lights. "Martin, you look a wreck."

"Sorry, dozed off – still a bit confused about time zones. Thought I was over it."

He ushered his dad in. George Truscott was bigger than his son, though losing some of his build as people can with age. Martin turned on some lights. The place looked a whole lot more cheerful. "So what's this 'Truscott and Partners'?"

"Oh, the old man decided to retire, and offered me the business. I don't think he really needed the money, so it was a ridiculously low price. I couldn't refuse. I cut a couple of the other guys in later when we needed a bit more capital to handle some offshore work. Civil engineering is really big, what with the mining boom. Pity you weren't interested in that . . .

"What have *you* been doing?"

His dad had found a chair to sit on. Martin was still standing, heading towards the kitchen.

"I've been making TV documentaries. So far, segments for bigger shows, but I've got a pilot of something bigger. Waiting for finance to do the rest of it. Do you want to see? I'm a bit behind in the cooking."

"Yeah, that usually works better when you're awake. Which is about the limit of my knowledge of cooking."

Martin handed his dad a DVD with the final edit, and the older man took it to the TV. "I think I can handle this."

As the video was playing, Martin busied himself with trimming artichokes, making garlic butter and finding ingredients for a couple more courses – rice, broccoli, salmon. He set the artichokes in a steamer, and called out to his dad, "Artichokes in about fifteen minutes."

There was a vague sound of acknowledgment.

By the time the artichokes were done, the rice was well on the way, and Martin brought the artichokes over with melted garlic butter and a bowl for discards. The timing was good: the video was close to the end.

"So, what do you think?" he asked as the closing credits

appeared.

"Pretty good. I am not that convinced myself of the science, and this makes me want to find out more. I really like the way you ask the viewer to judge. Nice. But it leaves me a bit unclear on what's real and what's not."

"Well, it's incomplete. I need more stuff from the real scientists. I left a copy with a possible backer."

"I could help out a bit –"

"Dad, it's not just the money. I need someone with the connections to pitch it to the TV channels. Things are so competitive now, even ABC and SBS are hard to deal with."

"Yeah, well you look as if you could use some help. I've never seen you so skinny."

"I had to do this on a real low budget – my own camerawork mostly, staying in hostels, lousy cheap flights. But it was worth it. Had some fun too?"

"Fun?"

"Met a girl."

"About bloody time. What happened to your veggie girl-friend? What was her name? Judy?"

"Julie. Not veggie. Green. Didn't work out."

"Yeah, the way you talked about her, I thought this could only be a heartbreaker."

Martin looked annoyed. "Eat your artichoke. I left the outer leaves on."

Artichoke eating is pretty labour intensive, so that occupied them for a while. Pull a leaf off (actually a petal, for the pedants). Dip it in melted butter. Pull off the soft part in your teeth. Eventually, you get into the soft core, discard the hairy bits and eat the tender heart.

"So, son . . . how serious is this sea level rise stuff? You've had it from the horse's mouth."

"Do you have a flood map of Brisbane?"

"Of course. We look at that stuff all the time. I have a copy of the 1893 flood map on my office wall."

"1893?"

"The biggest flood since Brisbane was settled. Eight metres above average sea level, 6.5 above the highest tidal level. Then there was the 1974 flood. Not quite as bad."

"And nothing since?"

"Don't you remember a flood in the 1980s?"

"Dad, I was pretty small then – not exactly reading the papers."

"You would have known. See, the Wivenhoe dam had been built by then for flood mitigation."

"I remember reading about that somewhere."

"Yeah. So a big flood today would be contained in the dam, as happened in the eighties. Hardly anyone even knows that one happened. Trouble is, all the flood planning assumes the flood will be from upstream, not from the sea. If we get a five metre sea level rise, it will be pretty close to the level of 1893 at highest tides."

"If I had any money to invest, I'd buy up land on the high spots, Dad. . . Why not choose some music?" He handed his dad his iPod, and pointed to the computer. "I've got some of that seventies stuff you used to like." He showed his dad how to set up a playlist on the computer. "When you're done, connect up the iPod, then click here when it's done synching, unplug it, and apply that engineering brain to figuring out how to play this over there." He pointed to the audio system at the other side of the room.

"I've got cooking to do."

He left his dad to choose the music, and went back to attend to the salmon and the vegetables. The rice was meanwhile nearly done.

Fifteen minutes later, he was back with two plates. "Sorry about no frills – I would usually do a cream sauce with this, and it should really include wine."

His dad took his plate and sampled. "Never mind. It's bloody good. Anyway I have to watch my cholesterol. Did you ever think of starting a restaurant?"

"No. You need a lot of money to do that, and it's not all cooking. A pretty tough business... Should you have had that butter with your artichokes?"

"Not really – but garlic is supposed to be good, so it cancels a bit. So is making TV documentaries easy?"

"No. But it's what I do."

"And if it isn't a success? Like computer games?"

"Dad, it's not like I have no useful skills at all. Do you remember that 3D graphics stuff from the New South Wales bunch? Where they showed changes when they varied the IPCC scenarios?"

"Yeah, pretty slick."

"They'd been working on that for months, and it was pretty slow. I fixed the main performance bugs in a couple of hours. Wilkinson has invited me to do a PhD with him."

"So that game stuff wasn't all useless after all. Will you do it?"

"Don't know."

"Tell me about this girl."

"Angie. She's studying biology at Berkeley, met her in Boston. She was checking out MIT for PhD options. She's really bright. We talk several times a day on Skype."

His dad looked sad. "When you talk like that, you remind me of how it was with your mum when we met. Whatever you do, son, don't have a disaster like I did."

"Yes. That's something I think we should talk about. What exactly happened with her?"

"I told you –"

"You told me lots of things that didn't add up."

"There were some things I didn't want you to know."

"I am old enough now, dad. What *happened*?"

The older man looked tired, and shrunken.

"So that's what it's all about. You have stayed away from me for so long because of this, haven't you?"

"Yes. Why do you think I was so obsessive, into computer gaming like it was a drug? Someone or something took my mother away, and you couldn't tell me the same story twice. Then, when I was getting to the point of asking her, she was gone."

"Can you imagine what it might have been like for me? The love of my life gone..."

"It may surprise you, but I can imagine."

"Imagining is not experiencing."

"You don't have to tell me."

"So you've had a bad experience too. The veggie girl? Why didn't you tell me?"

"We were not exactly on speaking terms, if you remember. You slagging me off for gaming. Me getting sullen about being lied to about mum..."

"So you saw it as being lied to?"

"What else was it?"

"Protecting you... I see now I've been a fool. I should have done more to understand. I thought it was just the usual sullen teen thing."

"Dad, I was still sullen long after you could get away with calling me a teen."

"You're right. You are old enough to know. But can you give your old man a bit of time to think it through, think how best to tell you? I haven't had to think about this stuff for ten years, and it's still hard."

Martin went over to his dad. "You know, we were never much of a huggy family, but I think you need one." His dad stood up. They did an awkward approximation to a bear hug – his dad stiff and uncomfortable.

"You know, Martin, you're not the same kind of idiot as your old man.

"Tell you what. I'll take you to a good restaurant tomorrow night, then we can go home, and I promise I'll do my best to make things right with you.

"OK?"

"OK, dad. Where should we meet?"

"I have a few options for restaurants. Let's meet at home at seven, then go from there. Give me time to choose one. You remember how to find the house?"

"Of course – if it's the same one."

"Yes. Graceville. I've done a few renovations, but it is mostly as you'd remember it."

It was still early when he showed his dad out – before nine – and felt wide awake thanks to the stupidly mistimed sleep. He washed the dishes, then sat contemplating for a bit, recycling the 1970s selection his dad had set up on the iPod.

Finally he decided: a walk would be good. He headed out, and found himself drifting towards the city centre. He ended up at Riverside by about 10:30, and contemplated the river for a while, watching the boats – ferries and party boats of various descriptions. The lights on the water played games with his eyes – he kept seeing images which he realized were flashbacks to the movie edits, not so long before, done in a building right behind him.

After about fifteen minutes, he said, "Goodbye river," as if talking to an old friend, and started walking back, this time a more direct route. It was close to midnight by the time he closed the door behind him, still not entirely tired. He undressed, took a shower, and lay in bed, contemplating. He had finally dozed off, when the phone rang.

"What the fuck, who could be calling at this time?" His immediate thought was Angie, confused about the time zone, but it was Prentice. Someone less likely to be riled by slips into rough language, he reflected, as Prentice began to talk.

"I was working pretty late today, didn't really feel like doing anything more, but just before I turned in, I thought I should check yours out. I at least owed you that . . . watched it four times . . . what time can you make it to the office tomorrow?"

"Do you mean tomorrow, or later this morning?" He found a piece of paper and a pen in the dark, and tried to start writing *Prentice appointment* but realized that he wasn't going to be able to write anything legible.

"Jesus F Christ. I didn't realize it was that late."

"That's OK – what time do you open? I have no other plans."

"Usually nine, but make it ten this time. See you in a few hours."

Prentice dropped the phone pretty hard.

Martin climbed out of bed, still a bit disoriented, half fearing that it was a dream. He turned a light on. But there it was: the piece of paper with the almost decipherable scrawl.

He went over to the computer. It was 1:30 am, 8:30 am the previous day in California. Angie would be awake. And there she was, on line. He started a call.

"What's this, a sex chat line?" She grinned at him. He suddenly realized he didn't have any clothes on.

"So sorry, is anyone else there?"

"Only my parents."

He ran off camera, and quickly pulled on a T shirt and some shorts.

"Am I respectable enough now?"

"I didn't say my parents were in the *room*. And anyway, how much do you think I can see of you – not much below the collarbone. Until you got up and ran."

"Ah crap, you really had me there."

"What time is it there? You've never called this kind of time before."

"1:30 am."

"What's happening? Or just can't sleep in your time zone yet? I thought you were just about over that."

"Prentice just called. He watched the pilot four times. Got so carried away, he didn't realize what time it was."

"Wow. Lucky you caught us all home – we're usually up by 6:30, out of the house by 7:30."

There was a muffled voice from her side. "That's my mom, she's outside the door. Would you like to meet her?"

"With my clothes now on, yes."

Angie disappeared from the screen, and a few seconds later, an older version of her appeared. "So this is the guy my daughter can't stop talking about."

"I hope so. I mean, I don't mind if she talks about something else ... otherwise it would be pretty boring."

She laughed. "She's right, you really are funny. In a nice way," she added.

"Now, listen. This is the best daughter I've got, and not only because she's the only one. If the two of you need help making a plan to be together" – offstage left, he heard Angie:

"Mom, no!"

Her mother carried on. "She's applied to MIT as you know. I've been thinking though, don't you have some pretty good universities in Australia?"

"I suppose. But she can't give up MIT on account of me."

"Everyone knows MIT is a crap shoot. They get so many straight-A students, they have to turn away a lot of the best of the best."

"I think she should talk to me about this, don't you?"

"Of course. I just thought you should both know, her dad and I will pay for her to go to Australia, if that's what both of you want."

Angie reappeared on screen, hugging her mother. "Did I ever tell you I had the best mom in the world?"

"Yes, darling, lots of times," her mother said.

"No, idiot, I was asking *him*."

Martin smiled. "I think I remember something like that. I'd definitely give her a positive vote."

Angie's mother gave him a smile, so like her daughter's, maybe lacking a little in impishness, and disappeared off screen.

There was a silence until there was the sound of the door closing.

"Just promise me you won't give up MIT for me."

"Promise me if I go to MIT, you will find a way to get to Boston."

"I promise I'll do what it takes. Right now, the only certainties in my life are around getting this movie together, and that should include a trip to California. I wish I had an option to go to somewhere like MIT. I had such lousy grades my last few years. I really do think I could perform if I got in. The New South Wales people were pretty impressed with the way I fixed their 3D model."

"So you could do a PhD there?"

"Yeah, but you wouldn't be there if you got into MIT."

"You know what I think?"

"You tell me."

"You are going to be dead on your feet by the time you get to see Prentice if you don't get back to sleep."

"I already slept a few hours this afternoon. Prentice will be in worse shape than me." He told her about falling asleep, and his dad's visit.

"You didn't tell me about your parents yet."

"My mum died when I was fifteen."

"Oh, that's so sad."

"You have no idea. My parents were separated before I was born, and my dad has never really told me what happened between them. I finally got to the point today of telling him he had to break it to me. He's going to tell me the story tomorrow."

"Poor kid."

"Listen to you talk. You must be what five years younger than me. *You* certainly aren't my mother." He grinned.

"I was just thinking about how it must have been – the gaming . . . was that coping?"

"More like compensation, no . . . hiding from reality. But that's all past. This should have been a rough week, confronting my dad, getting things on a new level with Julie, uncertainties of selling the movie, but I feel amazingly good. Something has changed."

"Whatever it is, don't change it back. I like the way it is now. But I am serious. You should try to get some sleep."

"OK, I'll do my best."

He lay on his bed mostly awake, but eventually drifted off, it seemed just in time to be woken by his alarm.

*   *   *   *   *

As Martin had predicted, Prentice was looking a bit the worse for wear. "I don't see any problem selling this. You exceeded my expectations. What would you like to add to complete the project?"

"I'd like to make it up to 45 minutes, with an alternative edit of an hour and a half. The main thing missing from my plan is mainstream science from some of the top places

– at least from California, also some in the UK. Good people in other places, but the MIT and Harvard people weren't interested first time, and at least with some friends in Berkeley, I have a foot in the door." Prentice raised his eyebrows, but Martin didn't explain. "A trip to the Antarctic may be an option, but I am not so optimistic about swinging that – I researched it, but the timing may be too hard."

"Have you got some costing?"

"Here you go." He handed two pages over.

Prentice shook his head again. "Still no crew."

"The way I did it with the handcam has a specific look that I think works – looks less slick, more sincere. And anyway, if I am going to reuse any of the original material, the new stuff has to be in similar style."

"OK. But I think we should make allowance for staying somewhere decent this time. Thirty bucks a day is ridiculous. Did you really come out on that?"

"Sort of ..."

"Sort of my ass. I can count your ribs through your shirt. Let's make it five hundred a day." He crossed out a few lines and pencilled in new numbers, doing deft mental arithmetic.

"What about editing time?" Prentice asked.

"If you look towards the end, you'll see I allowed for buying some equipment."

"Oh yes... And you forgot to pay yourself anything."

"What's the five hundred a day for?"

"Subsistence. You always expect extra when you are away from home. And how were you planning on getting to the Antarctic?"

"I have an invitation from the British Antarctic Survey.

But I would have to work out some academic connection, go as part of a project. I talked to Wilkinson at UNSW about it and he could swing it, but I think it'll take too long. If I do go, it would be on someone's research budget."

"Hmm. Better make sure this doesn't cause accusations of bias, if you get it together..."

"No problem there. I'll mention it in the final edit, so it's all up front and transparent. But more likely, I'll end up doing some phone interviews and patching in stock footage – can you help source some for me?"

"No worries." Prentice nodded. "Meanwhile, I've asked some of my connections at the ABC to view your pilot this afternoon. I'll let you know how it goes. I'm pretty sure they'll like it but if not I have a few other options. Meanwhile, I think you should take this contract home to peruse, and bring it back signed if you are happy with it. I'll let you know how the ABC goes, then we can make a time to meet again. OK?"

"So the contract is dependent on whether the ABC says yes or no?"

"More or less – but I'm pretty sure I'll find a buyer. Don't worry about that. I'll fund it anyway, if they haven't signed by the time you need to go. Just read the damn contract. Give it a day or so, take it to a lawyer if you like. Give me a minute, and I'll add in your budget." He typed in a few numbers. "I just put in the subtotals – travel, subsistence per diem, daily fee, editing equipment. When you look at your contract, you'll see these will come off your advance, after I find a buyer."

"And if you don't?"

"My risk. That's why I get a big fat cut and you do the hard work. But let's get word from the ABC before you start since they're likely to go for it – then we know where we stand."

"Sounds fair. Thanks. I'll read the whole thing. Wait for your call again?"

"Yes."

"I'll set my alarm for 1:30 am."

"Smartass." Prentice grinned. "Now get out, and let me get on with my other work."

*   *   *   *   *

The phone rang.

Martin picked it up expectantly.

It was Prentice again. "You'll notice it's 1:30 again."

"But pm – thanks for the consideration."

"Don't mention it."

"The ABC guy took a look, and said he'll take it to management. I have to write a proposal to make it official but you can take it from me, if he takes it to management, it's as good as sold. I've never seen him decide so fast. I gave it to him at 12:30, he called back at 1:15. Get me that contract, and we will have to wait a few days for things to fall into place. Don't do any shopping yet, because I could be wrong, but I don't expect to be."

"I'll drop it off in a couple of hours, if that's OK."

"No problem – if you're happy you had enough time to check the detail. If I'm out, my PA will be expecting you."

"Great. Thanks again for everything."

"It's nothing. You did all the hard work. I know what that means."

Martin checked the time. About 8:30 pm in California. For once, Angie was off line. He sent an email instead.

```
guessed right.  prentice was bit unsteady
on his feet.  but wait for it...  he's
buying it.  almost 100% certain abc will
take it.  that's our abc, not us version.
0 Crap, forgot the shift key.  Love you.
```

The rest of the afternoon was a bit of a haze. All he remembered was dropping the contract off at Prentice's office. *I hope I understood it. I think I get to keep some of the money...* Somehow the details didn't seem too important.

*   *   *   *   *

5 pm. Things were back in focus. He was going to see his dad that night.

Getting there required a bit of public transport research. He opened the Transinfo web site, and found some options. *Brisbane bus times are... approximate. Connections can be a nightmare – a scheduled bus doesn't show up, or is full and doesn't stop. So you start out half an hour early.* He decided to catch a train instead – the walk either end would be longer, but at least once you were on the right line, there weren't too many bungles possible.

His dad lived in Graceville, a leafy suburb to the west of the city. He lived a good few blocks from the train station. Martin got there with about an hour to spare, and decided to wander around the old neighbourhood. It brought back memories – some happy, but mostly just memories. He'd been such a misery as a kid. Then, at uni, he'd jumped on the train in bad weather and caught a bus for the last stretch, or ridden his bike on a good day. He didn't spend that much time in the streets near home.

There were some shops, a movie theatre, a lot of houses in traditional Queensland style – built on stilts, with wooden construction. Some were on quite a grand scale as Queenslanders go, others more modest. Most had reasonably tended gardens, a sign of owner-dwellers, not renters.

Even with the slow trains and the meandering, he had time to kill when he arrived at the house. It was starting to get dark, even though it was barely past six – even in mid summer, it doesn't stay light late in Brisbane. He walked in through the front gate. The house was dark; maybe dad was working late. He found his way to the back, to the old mango tree – still there, and hardly changed in all those years.

He could barely see the house from there, even though it was built high on stilts, the underside containing a garage, laundry and storage space, at least when he'd lived there. Who knows what could have changed. There was a big rain-water tank for one thing...

He sat on a patch of grass, remembering a small boy who had felt *different* from the start – parents who didn't talk, mysteries around his mother. In the fading light, he could almost see that boy sitting right where he was, talking to birds in gentle whistles. Of course, they did not speak his language and he did not speak theirs, if there was such a thing. But you could build a bond of trust if you were patient. A bird would hop still closer to you if you did nothing to startle it. If you made soothing noises, it might hop closer. Doing nothing was the best trick: slow movements, maybe. One day, a bird was close enough to touch. He reached out to touch its delicate feathers, and it was gone, so fast he didn't see it take off.

It was a long time before the bird (or maybe another like it) could be enticed anywhere near as close.

Trust. So hard to win, so easy to break.

He sat there thinking in the gathering gloom, the familiar scene somehow a comfort, despite the unhappiness that had followed that little boy's growth... the increasing distance of his mother, the increasingly unbelievable stories from his dad, her sudden, unexplained death, the retreat into extreme gaming, the roller-coaster of academic life – doing unexpectedly well here, unexpectedly badly there.

He had learnt about trust the same way as that little bird: it was so hard to forget when trust broke. But he wasn't a bird. He could learn. He could put this behind him. If his dad was ready to be truthful...

He was sitting there, oblivious of the passage of time, when a car's lights came flooding through the house, as his dad drove into the under-house space. He jumped up, as if to avoid being run over but of course the car stopped under the house; his dad obviously saw him through the shrubs, and yelled out, as he got out of the car, "Who's there?"

Martin waved. "Sorry dad. I didn't mean to surprise you. I got here early, and I was sort of daydreaming."

His dad beckoned to him, and the two of them climbed the stairs to the front door.

"I'm not used to having someone to greet me at home," his dad said, as he unlocked the door. "You know, I'm a goose. The better food places are mostly in the city, and you live the other side of town. We should have met in the city."

"I don't mind too much – I wanted to see the place and I got here just before dark. I wandered around the neighbourhood a bit too. Have you decided where to go?"

His dad meanwhile had hit some light switches.

"Yes – there's a new place in Toowong where I've entertained some big clients. Very good food. Have to match what you can do."

"Sounds good – and news on the movie: my backer is pretty sure the ABC is going to buy it."

"Great, son. I knew they'd go for it. When do you hear definitely?"

"I'm not sure, I delivered the contract this afternoon."

"So we can have a celebration – as well as the other thing. But first, let's look at the house."

They went from room to room. Much was as before, just newer. The bathrooms had been made over, and the kitchen was entirely different.

"Wow, dad have you taken up cooking too?"

"Actually, no – but everyone tells me the two things worth spending money on in a house are the bathrooms and kitchen.

"So what do you think?"

"Good job. Tasteful, still has the traditional feel."

His dad checked the time. "Still a few minutes before we have to go. Like a drink before we set out? You aren't driving tonight."

"No thanks." He remembered he'd forgotten lunch. "A drink without food will put me to sleep." He looked at the hall table, and spotted some real estate brochures, and business cards. "What's this?"

"You remember you asked me about flood levels?"

"Oh, yes."

"I checked the 1893 levels at the office."

"And this place...?"

"...would be under water.

"Do you think it would be something like insider trading to sell now?"

"Dad, I don't know for sure we are going to get multi-metre sea level rises, only that it is more likely than most people realize. But if it was my home..."

"...it is in a way isn't it?"

"I don't know. I haven't lived here for years. If it was mine, I would sell, just to be on the safe side." He looked again at the pile of real estate materials. "You move fast. I only mentioned this last night."

"Naah. I called in all these people to do valuations before the renovations. I just pulled all this stuff out this morning before going out to the office, when I was thinking about flood levels."

They drove to the restaurant almost in silence, each keeping their thoughts to themselves.

It was a new place, converted from an old house. The walls were vertical joinery, VJs, unpainted. Martin pointed this out to his dad. "Yes, I wondered about that too. I thought it must have cost them a fortune to strip off the paint. I asked the owner. This is how it was originally. Many houses of that era didn't have painted interiors to save cost. Looks quite nice. I thought of doing it at home with the other renovations, but it *would* have cost a fortune to strip off the paint."

The menus arrived.

They read for a while in silence, while tall arrangements of food precariously balanced on wide plates arrived at other tables.

"You know, I always think it's a waste to pay restaurant prices for oysters when you can buy them for less than a buck apiece at a supermarket."

"Son, do your old man a big favour."

"Yes?"

"Don't look at the damn prices."

They discussed options, then Martin's dad said, "What about wine? I could have a glass or two and still be safe to drive."

"Good idea – but you choose. I don't have much of a range."

"How about a Cloudy Bay Sauvignon Blanc?" Truscott senior pointed at a spot on the menu.

"New Zealand?"

"That's right. Most of Australia is too warm for a really good sauvignon. Tasmania is OK, one or two others, but I've always liked Cloudy Bay."

He ordered the wine, and they carried on reading, while the bottle appeared with an ice bucket and large, thin glasses.

Martin pinged his with a finger tip. "Classy. Not like the tumblers we get in hostels." He thought of the magical weekend with Angie in Boston. His dad's eyebrows moved, but he didn't request clarification.

A waiter arrived to take orders. Martin was a bit out of his depth, and his choices reflected something he could have done at home, to judge how good the restaurant was by a familiar standard. Oysters to start, barramundi and kipfler potatoes with lemon myrtle sauce to go for a local taste. His dad went for a starter of carpaccio of kangaroo, followed by Moroccan lamb with couscous. Both decided to leave dessert choices to later.

Seeing his son's slightly disapproving expression as his flattened out raw kangaroo arrived, the older man said, "That veggie girlfriend of yours trained you not to like red meat so much?"

"Green. No, she didn't. I went vegetarian for a while, true – she may have had something to do with that; we did a calculation of how much energy it takes to grow a cow, versus growing plants for food. But I kind of got stuck on the question of whether a lobster is a fruit or a vegetable."

His dad laughed. "I hope they didn't calculate based on the way the Americans fatten up their cattle on a diet of corn and soy. That's a ridiculous waste. At least ours mainly eat grass."

Martin ignored the riposte and touched his dad's gesticulating hand gently. "But dad, please. She has a name. *Julie*. Maybe I'm not seeing her anymore, but she meant a lot to me. And guess what? She has a respectable job in advertising, and is married, with a kid. She's more mainstream than me. You know, she helped me edit the movie, and it almost felt possible for us to be friends. I hope we can get there."

His dad looked awkward. "I guess I am the least qualified to advise on love, but the way you used to talk about her, it had to end in heartbreak." He realized he was repeating himself.

There was a silence.

"It did, didn't it?" his dad eventually added.

"You have no idea."

"Maybe I do."

The fish and lamb arrived.

"Son, let's talk about that crap later. This part of the evening's fun." He raised his glass. They clinked glasses. "Of course the wine choice is strictly wrong for what I'm eating, but if you can be unconventional, so can I. Anyway, you're the guest, and it goes with your order."

Martin smiled. His dad was all right when you stripped away the conservative engineer persona. He thought about his old persona – the game geek. Everyone can change. The old dog would learn new tricks, if someone took the trouble to teach him. As he started on the barramundi, he thought of what Julie had said about that persona. "Raccoon."

"What?" his dad looked up, puzzled.

"I was thinking of the editing session with Julie... she told me about when we first met, she thought I looked like a raccoon – pale skin, dark rings around the eyes..."

His dad couldn't stop laughing for several minutes. "She had you to a T. You know, I'm sorry now that that didn't work out. It couldn't have been so terrible if she's putting all this time into helping you."

As the main courses disappeared, the conversation slowed. Inevitably, this was leading to the Moment of Truth, as Martin was starting to see it. They decided against dessert and coffee, and headed for the car. To break the silence, Martin said, "You were right, that wine was a great choice. I hope it wasn't too bad with your food."

"Surprisingly good, in fact. You taught me something new."

"Unintentionally.

"The food was really good too; I must try out some of the ideas myself. I've never used native herbs."

They drove back in silence. The river was peacefully glowing in city lights, with the occasional boat rippling the surface. Then they swung away from the river between shops – a big multistory mall, car dealers, smaller shops. It all went by as it had so many times when each had done this journey before: Martin years ago, his dad almost every day.

Finally they were at the house.

With a few lights on, the place looked inviting.

"Sit down – I can get you a coffee if you like."

"No, dad. Let's get straight to it."

His dad drew a deep breath, and sat down. "You know that your mother walked out on me."

"Not exactly *knew* but it seemed likely. Did she meet someone else? What was he like? I never met him."

"It wasn't a 'him'. It was a 'her'."

"Oh."

There was a silence.

"But dad, why is this worse? If it was another guy, I could see you would feel it was an attack on your manhood or something... I couldn't look at Julie's husband, even though I knew long ago it was finished."

"What do you mean?"

"People don't choose sexual orientation. I mean, did you wake up one day when you hit puberty and decide, OK from now on, I'm going to be heterosexual?"

"But I thought we were so in love. When you were born, it was the greatest moment of our lives. You can't fake that."

"Maybe she did really love you. But this is something people can't control. Maybe she was in denial for a long time."

"I don't know. You have to understand, I was from a very conservative family. Very traditional. Very church every Sunday. The Queensland you grew up in is very different to the one I grew up in. I had no one to turn to, no one to explain this all to me. I thought you would grow up a highly confused kid if your mum showed up with a sort of step-mum but at her side. We had incredible fights over this. You don't know, no idea..."

"Dad. I didn't know the whole story but I knew for sure you were fighting each other. You can't hide that sort of thing from someone you spend hours with every day."

"You were so young."

"But not stupid."

"No one ever said you were stupid."

"But there've been plenty of times I've been a fool."

"Runs in the family." His dad managed a small smile at that.

"But then she suddenly started to get unwell, and no one told me what was going on with that either." Martin turned to his father, alert for the old half-truths and dissembling, but this time, once more he got it straight.

"That was a surprise. It turned out that this partner of hers was HIV positive, and passed it on to her. Or they got it at the same time, I never had the whole story. Just the partner was sick first."

"Oh. That explains a lot. But isn't that unusual, I mean for females to pass it on to each other?"

"Very unusual. That's why they caught it so late. In those days, the drugs were very potent, like doing chemo, and not always effective. Her partner managed to hang on until the therapy got better, but she didn't."

"Dad, by the time she died, I was fifteen. I think I was old enough then to understand. Even if I was useless around girls, I knew about sex. This was my mother we're talking about."

"I know. It's one of these things where once you've made the mistake, you just keep digging in deeper. Eventually just explaining why I didn't explain became the problem. But it was such a difficult thing. I still loved her very much, but her behaviour was something I couldn't deal with."

"And you kept her away from her son for that?"

"I really thought it for the best. She did nothing to make it easier for me. We spent a fortune on lawyers between us."

"Dad, did you ever see the movie, *North Country*?"

"No. What was that about?"

"A bunch of women fighting for their rights on a mine somewhere in the US. Never mind the story line. The thing is Charlize Theron is playing the young mother who had a baby before she was married and has an abusive husband. Her dad says she's always been such a disgrace to the family. Her mum says, she only had a baby, she didn't rob a bank..."

"...Did mum rob a bank?"

"I suppose... not. You have no idea how many times I've gone over what happened, wondered what I did wrong. If what you say is true about the sexual orientation thing, I suppose I was really harsh to her. I wish I could undo that."

"Dad, there is no rewind button on life. For a long time,

I was running away from my own past. Then I looked back, and found that it was moving away just as fast whether I did anything or not. The actual thing I was running away from was myself. I only really realized this after I met Angie in Boston.

"You made some mistakes. We can only assume mum loved you even if things broke down for other reasons. She couldn't have wanted you to be miserable for so long over her. Even if she didn't love you, you are hardly punishing her by being miserable.

"Have you got any ice cream? She liked ice cream. One of the few things I really remember doing with her is going to South Bank and stopping at one of the ice cream places. Why don't we do something nice to say goodbye to her, and do something she would have liked?"

"No ice cream here, but I know a good dessert place. Let's go."

On the way, Martin asked, "This partner. Is she still alive?"

"I don't know."

"I think we should meet her."

"I'm not sure if I can handle that. But I can easily dig up her last address from the time your mother died. It will be on a dozen lawyers' letters."

Martin decided to leave it at that for now. This was a celebration of sorts, after all.

An hour later, they were back at the house. After some conversation, his dad said, "Son, it's getting late and it would be silly for you to go home in the dark when there's a bed for you here."

"Thanks, dad." He checked his watch. "Have you got Skype?"

"What?"

"Voice over IP – telephone thing you can do on the computer."

"Well, I've got a computer and Internet."

They inspected the facilities. "A Windows machine."

"What's wrong with that?"

"Nothing, well not that much. But Macs come with a video camera and built in mike so you can use them straight off. Have you got a mike?"

"I don't know – I just took it out of the box. I only really use it for email and stock prices."

After checking the hardware, Martin concluded that it was a few parts short of requirements.

"Why don't you just phone her?"

"Her?"

"Your girl in California. Who else would you be so keen to talk to?"

"Angie. Dad, you have to get better with names. I've only ever had two girlfriends. It's not as if it's a harem of hundreds."

"Sorry. But anyway, feel free to use the phone. I'm sure you have a lot to talk about. I'll go to bed in the meantime. Your old room is made up, and you know where everything is."

Martin checked his watch. 11 pm – about 6 am in California. "I never tried it this way before. I hope they aren't sleeping in late for a change. I'll have to look up the phone number." He waved at his dad as he went off towards his bedroom.

He used the computer to look up the number. Luckily there weren't many Greens in Berkeley.

Then, noting the time, he took a shower, and killed some time checking out some research on the net. Finally, he felt it was late enough – Angie said they got up at 6:30. His dad meantime appeared to have finished his bathroom routine, and was fetching something from the living room.

The phone was picked up on the sixth ring.

"Hello, Green household."

It sounded too old to be Angie.

"Is Angela there?"

"I think so. Who is this? Not Martin in Australia?"

"Yes, how did you know?"

"My daughter introduced us on Skype, not twenty-four hours ago."

"Oh, yes. Stupid of me, how could I forget? It's been a very big day."

"Here she is now."

"Marty! What's up? Is your computer broken?"

"No. I'm at my dad's. It's been a long day."

"Your dad? Is he there?"

Martin called him over as he was about to disappear into his bedroom.

"Hello, Angela?"

"Hi. You must be a pretty good dad."

"Why?"

"Get to know your son better and you'll find out."

"He hasn't told me a lot about you yet, but I hope you can visit here soon. I'd really like that."

Sensing his dad was running out of conversation, Martin took the phone back.

"Good night dad."

He was still talking when his dad turned out his light.

As the older man went to sleep, he heard traces of the conversation.

He half expected to find Martin still at the phone the following morning, but his son was sound asleep. He quietly made himself some toast, and had a glass of orange juice. Martin still had barely stirred. He wrote out a note, and put it in the kitchen, with a spare house key.

Martin woke up half an hour later, feeling disorientated. Everything was familiar in a strange way. His room, yet the colours were all wrong... and this wasn't *his* room from recent years. Then it all came back. He staggered out of bed, found his yesterday clothes, and went out to the kitchen.

"Dad, I –" he started, hoping his dad was somewhere in the house, then found the note.

As he read it, he found some orange juice and an apple.

```
Martin

I know you have your own place and are
fully independent and all that, but this
really is your home as much as mine. Keep
this key. Feel free to show up whenever
you like, or not, as suits you.
Keep in touch, and let me know how the
```

```
project goes.

Love Dad
```

*   *   *   *   *

Back home, Martin was at a bit of a loose end. Making the pilot was so much part of his life, he had nothing else to do.

What would I do if Angie was here? Swim! She likes swimming. Let's see. The uni has a pool.

He looked up transport options, and decided to walk to the George Street stop, where there was a direct bus every ten minutes. Have I even got togs? When did I last swim? He grabbed a towel, and went via Queen Street, looking for a sports shop.

He found one, and added a speedo and goggles to his possessions.

At the pool, he paid for access, changed and surveyed the swimmers. Some lanes were zoned for speed. Better go with slow first. His swimming was awkward but he made it to the other side, and looked around. Most of the other swimmers were way better.

Only one way to get better: watch the pros.

He got out and watched the "fast" lane for a while, then tried it again . . . a bit better.

After several attempts, he felt some progress was being made. But, obviously, practice was called for.

The next few days, he didn't have much else to do but eat, sleep, talk to Angie and swim. Each time, the swim was slightly better. By the weekend, he was doing ten lengths easily, if not spectacularly well. He sneaked in questions about swimming technique, until he suspected Angie was onto him, and stopped. Muscles were starting to show definition in new places. She wouldn't see on Skype. Save it for a surprise for when they next...

But that was the problem. Unless Prentice delivered, there may not *be* a next time.

Monday. Nearly a week since Prentice had promised the ABC would buy. Martin gave up waiting for the call, and showed up at Prentice's office without an appointment.

A stern-looking woman guarded his door. "Mr Prentice is busy."

"For how long?"

"The next hour."

"I'll wait."

"He has another appointment the other side of the city."

"I can go with him and talk on the way."

"You wait right here and I'll see whether he can sneak you into a gap. What was your name again?"

Shortly after, she emerged. "All right, he's agreed to put aside what he's doing and talk to you." She said this in a tone that hinted that this did not happen.

Martin thanked her profusely, and went inside.

"Marty! So good to see you! You probably think I'm bloody useless, not getting back to you. ABC's unusually slow. But look, I am so certain of this, and I don't want to waste any more of your time, so I'm giving you your advance. Expenses, as in your budget, plus cash to cover costs. Rounded up to $50k, if that's OK with you."

It was. Martin could only nod. He had never had so much money before – even if much of it was for expenses.

Prentice signed a cheque and handed it over. Martin glanced at the amount, and asked, "Is there any paperwork?"

"You need to keep a travel diary – take this example to work from. You'll need that when you do your taxes. Make sure you record every day you're away. You'll probably get most of it back on your taxes. Keep receipts."

Martin glanced at the cheque again, then focused on the signature area. "This looks like a personal cheque..."

"Don't you pay attention to anything I tell you, kid? My damn accountant never gets his act together. Don't worry, I'm just advancing this to the company. I'll get it back. Not to be rude or anything, but we've both got work to do. Is your schedule still good?"

"Yes, yes. I'll have to confirm dates and everything because I wasn't sure when I could go, but I am more keen than you can imagine to get started."

"Good boy. Show yourself out if my door dragon has abandoned her lair."

Prentice opened the door. She was there – but not nearly as ferocious seen from the boss's side of the door. Martin smiled at her on his way out, leaving slowly to maintain his dignity.

As soon as he was out of the door, he took off like a rocket. It would still be afternoon in California.

He flew into the apartment, ripped the computer open, and punched Angie's name on Skype. He waited impatiently for her to pick up – about two rings.

"You will never guess," he panted.

"Easy. You're in training for the olympics."

"Well, maybe. But look at this." He held the cheque up to

the camera.

"Am I reading that right – fifty thousand dollars from Prentice?"

"Yup. Aussie dollars, but still lots of money to me."

"How much US?"

"About $40k."

"That's enough to get you to California."

"Then some – but California is where I want to be."

12 California

THE FLIGHT TO SAN FRANCISCO was not quite as bad as the flight to Boston had been. At least, after the hell of LA, he only had one more sector. And he didn't have a stupidly set up interview with no recovery time, the day he arrived. As Martin cleared the arrival gate, he looked around to orientate himself, and was almost knocked off his feet as Angie charged him and grabbed him as if to stop him escaping the planetary gravity well.

"Wow. I didn't see that coming."

"You won't see this either." She almost suffocated him in an energetic kiss.

Somewhere a hand appeared. "Hi. I'm guessing you are Martin. Angela's dad."

"Professor Green." He took the hand and they exchanged vigorous grips.

"Bob."

"Let me look at you." Angie pushed him back, and held him at arms length.

"I thought something felt different. Are these real?" She pummeled his shoulders. "Not implants?"

"I took up swimming. I'm not terribly good at it, but it does seem to have added a bit to my shoulders."

Angie's dad broke in, "Let's get your bags – if that's not all you've brought." He indicated Martin's backpack. "We can talk while waiting for them to show up."

"I do have a couple to pick up."

They went down the escalator to baggage claim.

Bob Green was tall and slim, with dark hair, greying slightly around the edges. He had the look of a former athlete – quick reflexes, yet not tautly muscled. He was wearing a T-shirt, but had a jacket over his arm. As they were waiting, he said, "It doesn't get terribly cold here, but I hope you brought some waterproof outerwear. It rains a fair amount in October – gearing up for the winter rainy season."

"Yes, I checked `weather.com`. I brought some skiing clothes I used to use when I rode my bike in the rain."

"You don't ride your bike anymore?" Angie asked.

"When I moved to the city, I didn't have a big space, and it was a bit in the way, so I sold it."

Conversation drifted on, and he found himself battling to stay awake. Then he saw the first of his two suitcases. He pointed it out, and Bob grabbed it.

"Not traveling so light this time?" Angie obviously remembered his single backpack in Boston.

"No. I have a proper budget this time."

Once they had the second suitcase, and were heading out of the airport, Martin turned to Angie and asked, "Where's your mother?"

"She thought we would be up to finding you on our own, and stayed home to tidy up the guest room. She's been pretty busy this week, and left it to the last minute. Grant proposals, paper deadlines, you know ... " He didn't really, never having been in an academic family, but nodding was Martin's default action right now.

The road towards San Francisco from the airport was rather boring – typical industrial near-airport scenery.

Bob said, noting his disappointment, "Nothing too exciting to see on the trip home, unfortunately. We go north on the 280, cross the Bay Bridge, go through Oakland and a few blocks from there, we're home. I'm guessing you may be a bit tired for scenery, but tomorrow, if you're up to it, we could go to Golden Gate and a few other touristy areas of San Francisco."

"I'd like that." He turned back to Angie.

She looked happy. "I hope the two of you get along." She put an affectionate hand on her dad's and Martin's shoulders.

They finished the trip in silence, Martin obviously not in good shape to maintain a conversation.

The rest of the day was a bit of a blur. Angie's mother introduced herself, and he forgot her name. He was taken to his room, offered a shower, taken to watch some TV ... he managed to stay awake through a dinner that was probably good but he suspected his face collapsed onto the plate several times.

He remembered saying at least once that he should stay awake until at least nine otherwise he'd never adjust to the time zone, but had no idea when he was put to bed. He had a recollection of Angie's face, then he was awake again. It was dark. He checked his watch. 2 am – 7 pm back home.

He was wide awake, yet still tired. He fumbled for a light

switch, the unfamiliar operation confusing him for a moment – US light switches are upside down relative to Australian. "Or is it that we're upside down?" He laughed at the old Australia joke.

Next thing, there was a gentle tap at his door. It was Angie. "I thought I heard something. Are you OK?"

"Just the usual jet lag thing. Don't stay up on account of me."

"I was just going to the bathroom. I couldn't sleep, thinking of you next door, anyway." She looked a bit unsteady on her feet.

"Let me put you to bed, and help you get to sleep." She didn't protest.

They went to her room, and put the light off. He eventually tiptoed back to his room, his work done, and decided to unpack, to kill some time.

After a couple of hours, he lay down again with the lights off, and managed to doze off in time to be woken up by household noises. He still felt a bit wobbly, but pulled on some clothes, and went downstairs, to find the kitchen, where breakfast was starting.

"I'm sorry, I look like a train wreck."

Angie's mother said, "No you don't. You just look as if you could do with a shower and shave. We don't usually have time for big breakfasts, but we're doing one in your honour. Go to the bathroom, and we'll be ready when you get back down."

Angie went upstairs with him to show him the facilities. "Did you really put me to sleep last night? I'm not sure if it was a really nice dream."

"Huh. I should be the one who's in no state to tell reality from illusion. I hope you don't usually lose sleep over me."

"Not a lot. I was just so excited to have you back." She handed him a big fluffy towel. "There's shampoo in about a squillion different flavours in the shower. I'd better leave you to this. I've been assigned chopping chores." She grinned affectionately. "And don't worry, I explained the vegaquarian thing."

Half an hour later, he was back in the kitchen, feeling better. The cappuccino machine was steaming lustily. A giant omelet, various enticing colours showing through from the interior, was emerging from a pan, and toast was popping out of the toaster.

"You timed that well," Martin said, as he took his place at the table.

Breakfast was a livelier occasion than the previous night – at least as far as Martin could remember.

As they finished, he said, "If anyone told me anything important last night, better repeat it. I don't remember too much . . . let's see. Someone fetched me from the airport." He turned to Bob. "I think it was you." A pause. He turned dramatically to Angie. "And *you*."

"Hey, you couldn't forget *that*!"

"Oh yeah, I was trying to remember where I got those bruises."

They all laughed. "And your dad is called Bob, and you mother is . . . " he turned to the mother.

". . . called Angie's mother," her mother finished.

"Damn. Ploy didn't work. You'll have to tell me your name again."

She laughed. "Us Berkeley Greens are not easily fooled. I saw exactly where that was heading. Your face was just about hitting your chest when you showed up last night."

She extended her hand across the table. "Melissa. And if you forget again, just ask." He nodded, and took her hand. Before he released it, she added, "And I'll let you know what the forfeit is."

When the chuckles subsided, Bob intervened. "Look, the easiest thing is if we refer to each other by name a few times, so no one is shown up. Anyway we don't want to spend all day around the kitchen table. A sunny Sunday in San Francisco in October – who could want to waste that? How about we pack a picnic and go to Golden Gate Park, and if everyone is still awake after lunch, we can look at some of the other touristy things we only do with visitors?"

Martin nodded.

"Since he's the guest –" he added, then paused, "whatever Angela's boyfriend's name is should tell us what he likes."

Melissa Green kicked her husband under the table – but everyone else thought it was pretty funny.

*   *   *   *   *

It was starting to get dark by the time they had finished the last touristy thing, riding the cable car – a kind of tram powered by a continuously moving underground cable.

Bob said, "Since we are doing the tourist thing, how would you feel about dinner at Fisherman's Wharf? There are some quite acceptable restaurants there, and that would get us back to where the car is parked."

"Fine with me." Martin was tired, but the time out in the sunshine had done a lot for resetting his body clock.

"We've been walking all day, except the cable car ride, so let's catch a cab." Bob hailed one, and they were soon heading back towards the Golden Gate neighbourhood.

Bob paid the cabbie, and they walked around a bit, inspecting the restaurants.

There followed yet another meal that probably deserved better than Martin's vague recollection, and a car trip that was hazy on detail.

By the time they got home, it was after nine.

"Do you still think you should go to bed around nine?" Bob asked.

Martin looked puzzled.

Bob reminded him, "Yesterday you gave us a long theoretical exposition on why going to bed at nine was about the right formula for breaking jet lag."

"Oh."

Martin thought a bit. "I am pretty tired now, but I think I can be kept awake a bit longer. I think the nine thing is strictly for the case where you really can't keep awake."

"Ah, a further development of the theory. I await your follow-up publication with interest."

"*Dad!* That's cruel. You can see he's not entirely with it."

Martin faced her. "No need to defend me. I've got bigger shoulders than him."

They all laughed. Melissa added, "I think the two of you might want to see if there's something worth watching on TV, or otherwise entertain each other." She pushed Bob upstairs.

Left to themselves, they curled up on the couch together.

"Now I'm getting to know your parents, I can see how you got to be this way."

"What way?"

"Smart, funny, great company ..."

"I thought you said you were no good with girls?"

A couple of hours later, she had to admit defeat – he really was only good for falling asleep, so she dragged him upstairs. "My turn to put you to bed." She pulled off his clothes. "My. Swimming does suit you."

"Mmph," was about all he could manage.

"Sleep, then." She pulled the sheet over him and gave him a delicate kiss. His subtle response was the only hint that he wasn't totally out.

She tiptoed out, turning off the light.

*   *   *   *   *

He slept through the night, but woke up just before light, feeling much better.

With a bit of time to spare, he prowled around his room, thinking about what had to be done. Connections at Berkeley to follow up, appointments to keep. Stanford likewise. *I should ask the Greens what the options are to get to Palo Alto ... dammit, I have some real money – I'll rent a car.*

He fired up his computer. While it was booting, he rooted in its bag for its charger and the US plug adapter.

Google maps.

He typed in the house address, then asked for directions to Stanford. Google maps obliged. Then it occurred to him that he hadn't done anything to set up Internet access. He checked the wireless connection: it was on something called "Greens at home". He shook his head. In Australia, because of download caps and charges, everyone password protected their wireless networks. He wondered if the Greens had a firewall. Anyone could walk past in the street and borrow bandwidth – not a problem if it's not metered – but cracking into your computers was another matter.

He did a bit of searching of the local network, and found four computers. All had a few open ports – none serious but potential security holes. Not the sort of thing a casual cracker could exploit – and you'd have to be in the street to get into the wireless network. No point in being paranoid.

He rummaged through his file of papers, and found the list of names at Berkeley. "Let's see, now. Bob is a professor of economics. He probably won't know any of these. No harm in asking though. Appointment today with Gonzales, ecologist. I don't suppose Melissa will know him either."

He went through the list of definite appointments, worked out where he had gaps, and started looking up the other people to see how best to contact them. "Just as well I left tomorrow free. It will be great if I can make it through Gonzales without falling asleep ... Might as well check email."

Nothing much new: a routine message from Prentice. Paperwork still working through the ABC. Still no obstacles, nothing to worry about. An old one from Angie, that he'd missed after leaving home. "Damn that stupid machine in LA that ate my money and didn't work." Nothing important, it would just have been great to have something to lift his mood at that dreadful airport.

It was starting to get light. "Should have asked for a house key, so I could go for a walk."

He took a shower instead, taking his time. No one else appeared to be awake yet. He found a passable set of clothes. "No point in looking shabby for interviewing."

Wake-up chores completed, he looked around upstairs. Aside from the three bedrooms and the guest bathroom, there were two well-appointed offices, one with a Mac in it, the other with a Mac and a PC.

Upstairs options exhausted, he went downstairs, and explored the TV guide. It all looked pretty boring – though PBS had some BBC stuff on that could be good. He prowled around the kitchen, exploring the selection of food, implements and appliances. Nice stuff, top of the line. He remembered the cappuccino machine from yesterday. A big one, not that many steps below commercial. The stove had powerful-looking gas burners. There was a ceiling rack with pots dangling down – all restaurant-grade copper, with stainless steel lining.

He pulled one down to take a closer look. It was heavy. The lining, though scrubbed clean, showed a range of food stains indicating steady use. He wondered who the main cook was.

There was a noise behind him. It was Bob. "Melissa told me Angela said you were keen on cooking. That's a pretty indirect message. Is it true?"

"Nothing lost in the translation. I'm a bit out of practice, though... living alone..."

"Well, in this house, everyone takes turns, so feel free to take charge if you want to do something."

"Thanks. But my immediate task is to track down some

people at Berkeley. I already have some appointments lined up, and need to find buildings. Do you know Evan Gonzales? Ecologist? Strong views on climate change and mass extinctions? My first appointment is with him, today. He's pretty busy this week, so I decided to chance seeing him when I might still be jet-lagged. Tomorrow is a gap, so I can catch up on anything I've missed."

"I don't know him personally, but of course he is pretty well known. I can help you find him. But I have grad student meetings, lectures, ..." he waved his arms expansively. "We'll need to get started early so I can point you in the right direction before my first meeting. What time are you seeing him?"

"I thought I should allow some time to find my way around, so I made it for 2 pm."

"Oh well, that's not so bad. We can look him up in my office, and if it's not easy to find, I can help you figure it out on the campus map. We could meet up for lunch if you like."

"Sounds good."

Meanwhile there were increasing sounds from upstairs – bathroom noises and the like, and Angie appeared in the kitchen. "*There* you are."

"Sorry. Not eaten by the cookie monster. I woke up early and ran out of things to do upstairs."

"Hmph."

"I couldn't wake you early. You have classes today. Even if they are inessential electives."

"Right. I'll take you..."

"Oh. Your dad just offered..."

"That's OK, you can go with him. I have classes and

project meetings all day."

She started foraging for breakfast.

Bob looked around. "Who turned the thermostat down? It's distinctly frosty in here."

Angie was a picture of contrition. "I'm sorry. I'm being a brat. I had this guy to myself for the best weekend of my life, but I suppose I should get used to sharing. This is work, and I want it to be good.

"Coffee anyone?"

She turned on the cappuccino machine.

"Mmm," Martin was conscious of a delicate situation, "are you really busy *all* day?"

"Why?" She paused at the coffee grinder.

"I was hoping we could go for a swim together. I've been putting in some lengths, but my times are terrible. I was hoping you could give me some hints. . . "

"Swimming is part of 'busy' – I always try to fit in a swim between noon and one – is there a gap in *your* busy schedule then?"

"An hour to my 2 pm meeting, maybe hour and a half if I don't swim too long. Let's meet there anyway at twelve, and if there isn't enough time, I can save up for tomorrow. I'll work it all out with your dad once I see the layout."

"OK – I'll wait for you there. The pool's called Spieker, right next to Rec Sports. Dad knows where it is."

Then Melissa showed up, and said, "If you have any appointments today at Berkeley, I can try to show you around between meetings."

When the laughing died down, she said, "I never thought I could have a second career as a comedian. Then again, maybe not: knowing why something you said is funny is probably a prerequisite."

The day was off to a good start.

* * * * *

Bob showed him to his office. It didn't take long to print a campus map, and Bob helped him pinpoint the building where Gonzales's office was housed. Martin added: "I'll look for it and take a walk around to get a feel for the layout."

"OK, Marty, we both have a lot to do. Do you need help with anything else? I have a meeting soon, and I need to get stuff together."

"Thanks, Bob, I think I have everything I need, except we need to think about how to get home ... "

"Oh yes. We all likely will finish at different times. If we can't work out something by the time you'd like to go home, I can give you a house key and I think there's a bus, though it doesn't run late at night, so it isn't much use to me.

"And one more thing. We originally planned to get lunch, but that was before Angela took over your schedule. What would you like to do?"

"We're going for a swim just before lunch, and things may be a bit rushed but I would like to see you again before Gonzales if we can somehow make all that work, because I could leave my stuff here then."

"Right. How about I get the three of us some sandwiches, and meet you outside the pool, maybe –" he consulted his schedule – "12:45?"

"Perfect. Meanwhile I'll take my swimming things for a walk around the campus."

Martin and Angela met outside the pool as planned. He was there a few minutes early, but didn't have to wait long. They went inside to get changed.

She examined his physique critically in the bright sunshine. "Hmm. Not bad in a swimsuit. A few more months of swimming, and you could pass for a real athlete."

"Well, thanks. I don't suppose I have to tell you that you look great."

"Enough. We have some swimming to do. Did you read the rules for lane sharing?"

"Yes – same as at home, except to swim on the right instead of left."

"OK – let's see you do two laps."

He waited for a gap, then jumped in, and did what he thought was a passable attempt.

"How was that?"

She leaned down close to him. "I wouldn't say it was terrible, but only because I love you."

He splashed her.

"Hey! Constructive feedback. You have arms and legs all over the place. Do you have some sort of water spider in Australia you're mimicking?"

She jumped in next to him, and let a few other swimmers through.

"Try to aim for a more symmetric motion. Roll about your spine, breathe on alternating sides, if that helps you stay more symmetrical."

"I'll try."

"I'll follow, and check if you cut some time off – that's usually a good indication if the lesson stuck."

Back again, he asked, "How was that for time?"

"Better, though I had to dog-paddle to avoid swimming into you. And I don't know what you call that turn technique. Frog in a blender? Now watch this for time."

She selected a gap in the flow of swimmers giving her a bit of head room, and took off like a rocket, doing a smooth turn at the other end, and flying back, almost catching up with the person ahead who had looked pretty fast to Martin.

"How did *that* look for time?"

"I don't know," he said, and ducked a drenching. "I'm not sure what the world record is."

She gave him a few more hints, and watched critically, then demoed her own vastly superior speed and style again. Then she stopped, and said, "I would usually do a lot more, but you have a meeting to get to, and I thought we could meet the gang."

She jumped out in a single smooth move, and he climbed out a bit less elegantly. She led him over to a group of tanned, mildly-athletic looking students.

"Hi, guys, I'd like you all to meet Martin."

"Gale, Dale, Linda, Jeremy, Pete, Marissa."

"Nice buns," said Linda.

"Pleased to meet you too," said Martin.

"You said he was funny," one of them said. They all laughed.

Linda persisted. "But seriously Ange, you didn't tell us he was such a babe."

"Babe?"

Angie laughed. "Non-sexist language. Anyone with a good bod is a babe. And he isn't going red because of sunburn." They all laughed again.

Martin changed the subject. "Are you all biology majors?"

"No," said Angie."Linda and Peter are both from my CS classes. Dale is in film. I met him at a party, not in classes – he wants to know about your documentaries."

"The rest are in bio then?" The others nodded.

"My docos ... the early ones were short segments in a science show. ABC – sort of like the BBC in Australia, not your ABC."

"Yeah, but we want to know about the climate change one." Dale pressed the point.

"Well, I don't know I have much to tell. It's my biggest project but it's still a work in progress. I don't really have anything to impress you guys."

"You've got to be kidding," Gale said. "Ange says you solved a coding problem that a room full of PhDs couldn't handle."

"Well, it was just some performance tweaks, and most of the stuff I used is in courses you can do here ... maybe have already done if you're CS majors."

Angie feigned umbrage. "Guys, please. Respect my man's modesty." He wrapped a towel around himself with mock deliberation at that. They all laughed.

"You guys are a comic act," Dale said. "Forget your day jobs."

Angie took charge again. "Martin is only with us for a couple of weeks. I don't know if we can set up a party while he's here –" cries of "Yes!" – "but I want you all to know, when it comes to my generous nature, and love of sharing," dramatic pause – "sorry, but this one's mine. But we will definitely have a party to watch his movie when it's done ...Marty, will you send us a DVD before it goes to air so we can have our own world premiere?"

"Yeah," added Dale. "Or put it on iTunes." He whipped out a slinky iPhone. "We can all watch it at the beach." The rest of the gang pulled out iPhones, and clinked them, in a kind of geeky toast.

"Hey, Ange, weren't you getting an iPhone in Boston?"

"Pete, I am not a slave to fashion."

Martin meanwhile was in a bit of a daze, not quite sure of himself. The geeky kid who only went to parties for the free food... Surely this couldn't be the same person – the envy of this bunch of hyper-achiever kids? If only he'd been so confident of himself then... He broke out of it. That was past.

"Seriously, guys, Berkeley is a great place. I used to spend the half of my life that wasn't spent playing and coding games on your web site picking up hints. In my raccoon days."

"Raccoon?" Angie looked at him with professional curiosity.

He repeated Julie's description. They all thought it was pretty funny. "We see people like that in our labs every day," said Linda.

He looked at his watch. "Angie, your dad said he'd bring some lunch to the pool for us. Is that OK? I wasn't expecting we'd have so many friends."

"I'm cool," said Dale. "You guys leave us here." He looked sulky. The others joined in...a chorus of sulks. They couldn't keep it up for long.

Pete added, as they were walking off, "If you impressed Ange, to us you're a god."

"You've got great friends," Martin said, as they went to change. She nodded. He wondered if it was such a big deal for her. You only know it if you haven't had it, he thought. Lucky girl.

They met outside, and waited for Bob, who was late.

"Are you the only person in the gang without an iPhone?"

"My trusty Nokia is still good, and so's my iPod ..."

"But Peter said you were going to get one in Boston."

"Yeah, a present for myself, to psych up for the MIT experience but I got a better one." She hugged him. "Also, I bust my budget a tiny bit. That Pinot Noir ..."

"Not the price of an iPhone!" It was good, but not that good – or was it? He'd never had a wine costing more than twenty bucks or so, except the one he'd bought in Boston – that was about forty bucks. So I think she's special, she thinks I'm special.

"And those glasses ..." she added.

"Surely *not* as much as an iPhone!"

"Well, no. But a big enough dent in the budget that I started thinking...what if I did follow you to Australia? Won't you have a different phone system there?"

"You were thinking of that even then?"

Bob showed up with the sandwiches.

"Still over an hour to your meeting," he huffed, "sorry I'm a bit late. Did you find Gonzales's office OK when you went scouting?"

"Yes. Let's have lunch, then fetch my stuff. When's your next class, Angie?"

"Right after this – I'll have to run after eating this." She looked pleadingly at her dad. "Can I take him home at the end of the day?"

"Of course. Your mother and I both have to work late – grants, papers, yadda, yadda. I was hoping you could handle it."

"Can and want. Great, Marty, let's meet back here at five. My car isn't far from here."

They ate the sandwiches quickly, then Bob and Martin went back to Bob's office. Angie ran off, part of the energetic mob heading for classes.

With a little time to spare, Martin wandered slowly over to his destination, mulling over his prepared questions. Finally, he was at the right door. Evan Gonzales was clearly waiting for him.

"You must be Martin Truscott. Good to meet you. My friends at NASA say good things about you."

"Thanks. I like to film with natural light if I can – is there a spot outside that you like?"

"Yes – the botanical garden is not far off, and there's plenty of foliage out there to add colour. It is a bit of a walk, but I like some exercise if that's OK with you – if you don't have heavy gear to carry."

"Oh, no – just what I'm carrying."

They walked in silence, until they were almost there.

Martin asked, "Do you have any questions before we start? I think I made my intentions explicit in the email."

"No. Let's get straight into it. Will you let me check the

recording afterwards? Sometimes one says the wrong thing
..."

"No problem with that." Martin smiled. "I'll send you the
final edit before it goes to air as well, if you have any doubts."

"Thanks. It's good of you to offer. Here's a nice spot."

Martin set up the camera. "Yes, that looks good. Let's
start."

"Professor Gonzales ... you're a professor of ecology at
the University of California, Berkeley, and one of the top authorities on habitat and endangered species. What is your take
on global warming?"

"Habitat destruction in any form is a huge problem. Once
you've removed the various synergies forming an ecosystem,
it may be gone forever."

Martin nodded. "What about this position some people
have, that the number of extinctions is in fact very low?"

"First, observed extinctions require that we know that a
species exists. Vast tracts of rain forests in the Amazon,
for example, have been cleared, without any scientific survey ever having been completed. Second, some of the numbers I've seen from climate change doubters are way below
numbers accepted in academic literature – up to ten times too
small." Gonzales waved his arms for emphasis.

"How much of this, though is directly attributable to climate change?" Martin asked.

"Of course clear cutting and burning in the Amazon is not
directly caused by climate change. However, we are starting
to see habitat destruction more directly attributable to climate
change. Polar bears are of course the iconic example of a
species increasingly under threat. They are highly dependent

on sea ice for feeding – stalking walruses and seals. They do very poorly if they have to hunt on land. The scenario of bears drowning because they have to swim long distances between ice is more dramatic perhaps, but not the biggest threat."

Gonzales paused to catch his breath, and Martin followed up: "What of this argument that there has always been climate fluctuation?"

Gonzales pursed his lips then enunciated clearly, as if explaining to a first year class. "If the IPCC scenarios are correct, and I have no reason to doubt them, the rapidity of change is a major issue. The date by which the north pole region will be completely ice-free in summer keeps moving forward. This alone will be an enormous shock to many ecosystems. In any case, it's silly to claim that because there have been previous climate change events, this one can't do any harm. Ask the woolly mammoth. Ask the neanderthal. The difference today is that the woolly mammoth and the neanderthal neither created nor had the power to stop climate change. We are guilty of both."

"Could you give us some numbers for the rates of extinction before and after industrialization?" Martin asked.

Gonzales nodded. "Yes. The generally accepted pre-industrial rate was about one in a million per year – this was the pre-human figure, and lasted pretty much until human power for destruction was extended by machinery and powerful weapons."

"And the figure today?"

"There is a range – as I said before, this is hard to measure – but anywhere from a hundred to a thousand times faster than the pre-human rate." Gonzales spread his hands, indicating a wide range.

"And with climate change?"

"It really depends how severe the effects are. We can expect animals at the extremes to suffer the most. Emperor penguins: emperor penguins are a case in point. They have an extremely specialized lifestyle adapted to breeding in the Antarctic. They are remarkable birds. Any significant breakup of the Antarctic ice sheet would severely jeopardize their chances of survival. A terrible loss. We don't expect as much temperature change in the tropics but the tropics and deserts are some of the least studied habitats. Many exotic species could disappear."

"What of the people who say this is not so significant, the figures are exaggerated, extinctions happen all the time?"

Gonzales looked annoyed. "They aren't in the field. Show me the work they've done. No serious ecologist isn't concerned. Most of the biologists I talk to are concerned."

"How does the current situation compare with major extinctions on a geological timescale?"

"Well, 99.9 percent of all species that ever existed are extinct. Not just individuals, whole branches of the evolutionary tree. Dinosaurs, large amphibians, giant marsupials for your Australian audience ... We have evidence for six great extinction events. Around 488-million years ago, 444 million years ago, ... more recently, the transition between the Cretaceous and Paleogene eras 65-million years ago. The end of the Cretacious was the end of the dinosaurs. Half of all species died off then.

"We think we are now in the midst of another great extinction event, the Holocene extinction event. Some call it the 'Anthropocene' extinction event because we are causing it. Many are of the view that up to half of all species could be gone in the next five hundred years – an extremely short period on the geological time scale."

After a brief pause for emphasis, Martin followed up: "Let me get this straight. You are saying this is as big as the end of the dinosaurs?"

"Yes. Potentially. We need to think through carefully what the dominant species are of our era, the ones most adapted to the way things were. In the Cretateous, it was the dinosaurs. I should make it clear what that fifty percent means. Not fifty percent evenly spread across the board. The plants and animals least adapted to change could go as a group. If for example a couple of our major food crops went, humanity could be in deep trouble." Gonzales's hand gestures conveyed his sense of urgency.

"Are you saying it could be *us* this time?"

Gonzales looked serious. "I am only saying it was the *dinosaurs* in the Cretaceous. And they were incredibly successful for over a hundred and sixty million years – much longer than humans have existed."

"Thank you professor."

Martin turned off the camera. "That was great. Can I get you a coffee or anything? I don't have anything else to do until about five."

Gonzales consulted his watch. "I don't have a lot of time. But why not? You can tell me who else you've been talking to."

Gonzales found a coffee shop, and they discussed the confected controversy principle by which the press confused the public, the NASA people's views, the value of having real scientists running `realclimate.org` and the question of the ethics of funding from the likes of tobacco and fossil fuel companies. Martin steered the conversation away from specifics of other interviews. "I don't want anyone biased by what they think everyone else is saying to me. I told one bunch what someone else said, and that may have been a mistake. Your views stand pretty well on their own."

Gonzales nodded. "Very professional. I know I said I had almost no free time the rest of your stay in California, but it's been a pleasure. Let me know if there's anything else you need while you are here."

The coffees finished, they parted, and Martin wandered around the campus for a while.

Martin wasn't much of a student of architecture, but he picked up Spanish and classical influences. He found a shady spot away from prying eyes, downloaded the interview, and copied it onto a couple of DVDs for safe-keeping. He decided not to replay it right there, good though a pick-up ploy that had turned out to be. "One like that is more than enough to keep me busy." He sat back and relaxed.

This was turning out to be a great trip.

The rest of the day should be quiet, and he had all tomorrow to set up questions for some of the more difficult interviews. Time to relax. He sat back and watched the students moving between classes. All so happy, energetic... could we really be the next dinosaurs?

*   *   *   *   *

They met outside the pool at five. Martin hardly waited a

minute before she showed up.

"Come on, let's get home. I want some time with you while you're awake."

He grinned, and sprinted after her to the parking lot.

She drove for a few minutes, before asking tentatively, "So what did you *really* think of my friends?"

Ah, so she wasn't totally self-confident. The next words came out without too much thought, as he contemplated this development. "It's been a while since I was a student. I wouldn't have thought a five year gap would be so great."

She concentrated on driving. He wondered if he'd said something wrong. He couldn't fathom her expression.

When they got home, she slammed her door, and locked the car as he got out. She rushed to the front door, let him in, and locked it behind him.

She grabbed him tightly. "So you think I'm just a kid? I'll show you..."

She dragged him, protesting, up to her room. She hurled him to the bed, as he was uselessly still trying to say, "I didn't say..." – then stopped protesting.

Finally, she said, "So you think I'm just a kid. Did anyone ever do *that* to you before?"

He took a while to answer. "No. But I *didn't* say I thought you were just a kid. You will *not* listen."

"Oh." She sat up and pouted. "I promise never to do that again."

"I most fervently hope you mean the bit about not listening."

"Do you think I didn't enjoy the other thing?"

"I really hope you did. I was just trying to say I could get back into being a student – it's not the five years that separates us. In some ways, *I* am still a kid – I know less about life in some ways than you do. The only reason not to go to New South Wales after all this is, if you go to MIT, I'll have to find a way to get to Boston."

She nodded. "Well, we should know soon. If you apply for February entry, they promise a quick response – most students apply for September. Anyway, the timing worked out pretty well – I wanted to be here when you were in California."

As they lay together contemplatively, the light faded.

Suddenly a car turned into the driveway.

"Oh, crap. Look at us." She threw his clothes at him. "You need a shower. We can't let them see us like this."

He caught them deftly. "Otherwise, shotgun?" He grinned.

She pushed him out the door. "Idiot, just get in the shower. I'll tell them you were exercising heavily."

"That's not a lie." The bathroom door closed behind him.

A minute later, another door opened, and Melissa's voice floated upstairs.

"Angie, there's something in the mail for you."

"I'll be down right away mom."

Ten minutes later, Martin emerged from the shower, respectable again. Angie invited him into her room. "Guess what?"

He shook his head.

"The thing in the mail. MIT." She watched him closely. "I'm not going. What do you think my chances are for New South Wales?"

"Well, if you are a straight-A student, pretty good, I'd think, but it's expensive for foreigners to do PhDs in Australia, probably about $20k a year in tuition alone. And hardly any scholarships for foreigners, not like here."

"Mom did promise. I think they can afford it.

"I think I should take a shower too." Another car arrived. "That will be dad. Why don't you talk to him while I clean myself up?"

Martin went downstairs.

The parents were engrossed in conversation.

Bob looked up. "Oh, hi Martin. I was wondering when you two love birds would deign to join us. One is better than none."

Martin looked embarassed, so Bob quickly went on: "Tell us about doing a PhD in Australia."

"Well, I didn't ever need to look into too much detail, but I suppose it's not much different from here, except you don't start out doing courses. You need a good 4-year degree to get in."

Bob nodded. "So far, except for the courses, same as here."

"Yeah, a lot of our degrees are only three years, like my BSc."

"OK, there's a difference. A good university would have a 4-year BS."

"So to get into a PhD I would have had to do an extra year, Honours – that's actually another degree."

"And did you think of doing that?"

"Well, maybe a bit. But I was depressed about my low grades over the last year, so I gave it up."

"Pity," said Melissa. "How did you get into documentaries?"

"I was doing whatever work I could, contracting, fixing computers. A Mac dealer sent me out to help someone at the ABC – that's like the BBC in Australia – set up a computer. He got talking about stuff like what was all this dual core everyone was talking about. Just conversation. I explained it to him. He said something like, 'No one ever explained computer stuff so clearly to me before – would you like to help out with a segment I'm putting together for *Catalyst*?'"

To their unspoken enquiry, he added, "Science show. Anyway, I ended up doing the whole thing, and that gave me the connections to do a few more. Trouble is, they didn't pay that well, if you took into account that the research time killed my other contracting work. I couldn't do it by half measures. I picked up that melatonin may not be such a wonder drug for curing jet lag, and I ended up traveling to three cities to interview medical profs to make a fifteen minute segment. The multicore one was the only one where I knew enough to do it without research."

"So here you are now," said Bob. "How good is the University of New South Wales? I think I've heard of it."

"UNSW is supposed to be one of the top universities in Australia. We have the 'sandstones' – mostly the traditional, older universities – and newer ones. I think they did the same thing in Britain: converted a bunch of technology colleges to universities. Anyway UNSW is supposed to be one of the better ones."

"I see," said Melissa. "The system here is very hierarchical. The California state system is nominally one big system, but it's highly filtered. Only the best of the best get to the top tier, like Berkeley."

Martin thought for a bit. "Yeah, that does make some sense, but in Australia it's not quite like that. Students get conned by marketing, or just go to the place closest to home, especially at undergrad. The difference is much bigger for PhDs. I was at the University of Queensland which has about the same ranking as UNSW, and the differences in my class-mates amazed me. We had some real dunces, and some students who were so bright, they'd fit right in here. Research is different. The top few including UQ and UNSW are off the scale compared with most of the rest – the top eight get something like seventy percent of all research funding. The PhD students I met at UNSW seemed pretty sharp – I just happened to have skills in a different area. I don't think I could have shown them up in anything else.

"Australia is really big in the bio sciences. I think some-one recently got a Nobel, and we have massive labs – world-class stuff. I wish the computer science was as well funded."

Melissa nodded. "So for a PhD, she'd probably do OK there. You see, it's so competitive here. She would miss the push to be the best of the best."

"In Australia, elitism isn't much tolerated. I suppose that homogenizes things a bit at undergrad. Research is more in-ternational – you have to compete with the rest of the world. People there have this tall poppy thing – stick your head up too high, and you get cut down."

"Don't they have sporting heroes there? I thought Aus-tralia was sports mad. And there was that crocodile guy, what's his name?" Bob looked unconvinced.

Martin nodded. "You're right – physical prowess seems to be good. A bit inconsistent."

"If you ask me," Melissa butted in, "it's not anti-elitism, it's anti-intellectualism."

"Maybe you're right. The conservative politicians at home have a way of sneering at anything vaguely intellectual, even beverage choices."

"Beverage choices?" asked Bob.

"Yeah – people who drink latte or chardonnay."

"Oh, yes – beveragism." They all laughed. "We get exactly that here too. But," Bob continued, "this is serious stuff – the new bigotry. You sneer at people who differ from you. You aren't allowed to sneer at blacks anymore, and not quite as much at gays, so you make up a way to differentiate people from the mass. Then marginalize their politics, because they aren't mainstream for whatever stupid reason."

"OK, like, the ordinary joe who swills beer and drives a ute..."

"Ute?" queried Bob.

"Uh, pickup. That's what we call them in Australia. The ordinary joe...   as if politicians drink cheap beer and drive pickups."

"Dog whistles," added Bob.

Martin nodded. "I've heard that one too – innocuous stuff with hidden meanings only some people pick up. Is there *anything* original in Australian politics?"

Melissa replied: "Tall poppies possibly – I haven't heard that one before. But anyway – I can just see the faces of those snobs at MIT when they get her 'no thank you'."

"Her what?"

"Didn't she tell you? She got an accept from MIT – that's what was in the mail – and decided straight off to turn it down."

Martin was taking the stairs up two at a time and almost cannoned into Angie as she emerged from the bathroom.

"We need to talk."

She was surprised at the look on his face. He steered her to his room.

"Sit there." He pointed at a chair, and sat on the bed.

"You promised me you would not turn down MIT on account of me."

"That's not exactly what happened. I have a very good memory. You said, 'Just promise me you won't give up MIT for me.' I said, 'Promise me if I go to MIT, you will find a way to get to Boston.' You didn't exactly promise that."

"You made it sound as if MIT had turned you down."

"I didn't exactly say that."

"Look: this is not a court of law. I want to be able to trust you. You know what I feel about being lied to."

"I didn't lie."

"That's not the point."

"What *is* the point?"

She stormed out, slamming the door behind her. He collapsed onto the bed.

Fifteen minutes or so later, Melissa knocked on the door. "Can I come in?"

"Yes," he said quietly.

She sat next to him, as he sat up. "You look miserable."

"I don't know how to deal with this," he said slowly.

"Why is this such a big thing? She wants to be with you. It may cost us a bit, but we can find the money."

"Yes, but she wasn't entirely truthful, and she knows I've had bad experiences with that."

"She is so competitive. Everything she has ever done before has been aimed at being the best at everything. This is the first time she's turned down something this big."

He smiled. "I saw that competitiveness when we went swimming. But why couldn't she tell me this straight out?"

"I don't know. I suppose she's always been able to control every aspect of her life before."

"But she doesn't need to control me. I don't want to control her. Things have been so good with us because no one is in charge."

"This is new terrain for her. Did you see her with her friends yet?"

"At the pool. She was obviously dominant – the alpha female."

"Exactly. Her friends are all the brightest of the bright, and she is a class above. She's never had a friend on equal terms before. It's a bit of an adjustment – but I'm so pleased to see it. I talked to her a bit, and I think she gets it. No more control games."

"It's great she has support. I'm all alone here..."

"No, you're not. She's *my* daughter but you can count on me if you need help. OK? Bob and I will stay out of the way if you need to talk."

"And she's at the top of everything she does. This doco is the first thing I've done where I haven't felt I'm a total loser."

She gave him a hug. "Don't you worry. She's had her

failures too. When you feel the moment's right, ask her about her attempt at breaking into music. Everything was lined up right except one thing. Tone deaf. Not only her, her buddies too. A total flop. But I didn't tell you that." She winked.

He nodded, his mood lifting. "And I guess that may explain why she is so competitive. Maybe compensating just a tad? How many boyfriends has she had before me?"

"I don't know – two, maybe three serious – countless admirers and dates."

"Just so you don't think I'm sophisticated and better able to cope – I only ever had one girlfriend before, and that left me in a bad shape for years."

She smiled. "These things don't always work out, but sometimes they do. If they don't, life goes on. Bob and I have been together since we met in grad school. We have some vicious verbal battles, but it's all very civilized. We work it all out, and whoever was wrong has to pay a forfeit. Usually it's both of us.

"Come on – get this one out of the way."

They walked out together. He found Angie on the couch downstairs, in front of the TV, obviously not focused. The lights were low. He touched her on the shoulder gently. "Let's talk," he said quietly.

She made space for him.

"You've been crying." He touched a wet patch on her face. He was genuinely surprised: she had never seemed the tearful type.

"Look what you've gone and done to me. No one has *ever* done this to me before."

"You did it to yourself."

She cuffed him. "Ouch. Damn those hard muscles."

He laughed, the tension broken, and pulled her close. He started talking softly, barely above a whisper.

"When I visited my dad's house, it brought back memories. I was sitting in the back yard under the old mango tree, waiting for him to get home, and I remembered a small boy who used to sit there – lonely because his dad worked long hours, his mum was a stranger, he didn't know how to make friends.

"This little boy used to talk to the birds. There were delicate little birds, black, with a long tail, inquisitive, hopping around happily. He would whistle little tunes at them, hoping they meant something in their language.

"Of course they didn't, but the birds got used to him, and hopped closer and closer.

"Then, one day, one of them was close enough to touch. He reached out a finger to touch the delicate, shiny feathers, but the bird was gone, faster than he could see, and the birds never got that close to him again."

"I thought *you* were a bird with a broken wing."

"That too," he grinned. "But the point of this little story is trust takes time to win, and can be lost really easily.

"I think I screwed this one up a bit by making it all or nothing, holding you to promises without options. It would be great if you could be perfect but then I'd have to be perfect too. Can we start again?"

"You mean another weekend in Boston?" The impish grin was back.

Dinner that night was a less festive affair than previous times.

Martin and Angie quite forgot to call her parents down when the crisis was over. Eventually Melissa peeked downstairs, saw they were laughing, and cleared her throat delicately.

"Oh, mom! I forgot we were supposed to tell you when we were OK."

"That's OK, dear. Is anyone hungry? It's getting late."

Everyone was a bit.

Bob was also on his way down, and was obviously a bit impatient. "Let's just send out for pizza then."

Twenty minutes later, they were downing slices of pizza.

Melissa asked, "How does this pizza compare with Australian?"

"Not bad compared with Dominoes, who are everywhere, but I like pizza in a more Italian style. Not to be critical or anything," he added quickly.

"No problem," Bob laughed. "Your forfeit will be to cook tomorrow night."

"Pizza my style?"

"Great."

As they finished, Angie said, "I have this essay to finish off – maybe thirty minutes. Creative writing. Can you all leave me alone for a bit?"

As she left, Melissa asked, "So, Martin, what do you do differently with pizza?"

He started to explain how he made a simple dough of yeast, water and flour, and kept it springy by working in flour when shaping it. "One thing I don't do is use a heavy cooked-down tomato layer. I start with mozarella, and put uncooked tomato on top – has a fresher taste than the usual sauce base."

An extended discussion of pizza theory followed, and thirty minutes went by fast.

Martin glanced at his watch, and said, "She must be about done by now?"

Melissa nodded. "She's actually pretty amazing. I never knew how long a piece of work would take – I still don't. Sometimes I am up all night even now finishing off. If she said thirty minutes, it *will* be just about done."

He went upstairs. Angie was sitting in her room, working on the computer. Martin walked in. "Still busy?"

"Just finishing off... Sit here. I like having you close but it would have been boring to watch me type for half an hour."

He sat next to her, watching her make corrections and adjust the layout.

"Done!" She turned to him. "Your turn to choose what to do."

"Let's talk about what we're going to do. Did you reply to MIT yet?"

"No. But I am *not* going there unless we can find a way to get you there."

"Hey, look what happened when I started out with non-negotiables...

"How about this: you tell MIT you'll go there if they can find something for me, and I'll tell Wilkinson I'll go to UNSW if he can get you into a program of your choice?"

"I don't know. MIT has all this bureaucracy around PhD applications..."

"I don't care about that: it doesn't have to be a PhD. I can work in a lab, drive garbage trucks, whatever."

"I don't like that idea. You are already playing in the big league."

"OK, let's give it a try anyway. UNSW seems to me a whole lot less bureaucratic. They can afford to be, I suppose – poor old MIT must be flooded with applications."

"I've still got my transcripts from my MIT application. We could scan them and email them to Wilkinson, if he can help us get a quick opinion. What kind of city is Sydney?"

"Bigger than San Francisco, smaller than LA. I only spent a few days there. I think we could get to like it. Same as we worked on Boston together, if you like ... "

"I like," she interrupted.

He decided to file the Boston trip away as the thing to mention whenever she was down. Then again, he also felt pretty good thinking about it. He continued: "... except this time I have the head start."

"I don't care, as long as it's a joint project."

He nodded. "Making a pitch to MIT will be the harder one – they don't know who I am, so let's work on that."

It was heading towards midnight, when they were happy with the letter.

"Time for bed," he said. "I'll just fire off a short email to Wilkinson. Dunno what he can find out by the time we wake up because it's almost 5 pm there, but give him time to think about it." They had a lingering kiss, then he went off to his room, and typed and sent the email.

```
I'm thinking seriously about the UNSW PhD.
Can you find out for me whether I could
get in, as discussed?  (Lousy grades, no
```

```
Honours year.)
```

```
Thanks.
```

He was about to close the computer, when a reply came in – almost as if Wilkinson had set up a reply-bot:

```
I'm way ahead of you.  It's cleared already.
If you have a degree -- any degree -- you're
in.  I have a scholarship saved for you
out of one my grants.  $20k pa tax free.
If you need more, I'll work on it.  When
can you start?
```

He rushed next door with the computer. Angie was undressing, the light still on.

"Wow. MIT had better hurry to get their act together," she said.

"Not so fast. We have a detail to take care of."

He typed a reply.

```
We have a 2-body problem.
```

"Two body problem?" She pointed at the text, then at each of them.

"Physics talk."

"Oh, right, gravity. This is a matter of great gravity."

He grinned and carried on typing.

```
I can't do it unless my partner Angela
Green can get into a bioinformatics PhD
in Sydney.  Straight A student from Berkeley,
```

```
accepted into MIT already.  Can you swing
it?  I can send scanned transcripts tomorrow.
```

"Partner?" she queried.

"I'm sorry, does it sound too formal to you? Like almost married?"

"No. It's just nice to see it for the first time. Send it."

He did. A few minutes later, Wilkinson replied.

```
I don't see how it could be a problem but
I'll have to ask a few of the bio people
once I get the transcripts.  The accept
letter from MIT should help.  Some hints
on what she wants to do as well.
```

"Really bed time now," he said. After the inevitable kiss, he added, "Partner."

*   *   *   *   *

The next day, Martin woke to daylight. The house was quiet. He pulled on some clothes, and looked out of his room. Angie's door was open and she was working at her computer.

"Oh, there you are. The others have gone already, but I'm skipping classes today – except the stuff I can do from home, which is everything except sitting in lectures and getting bored.

"What's your plan for today?"

"Well, I don't have any appointments today, but I should follow up on the ones who couldn't give me definite times, and do some more background reading. I also need to look into renting a car for Stanford."

"You mean you don't want me to take you?"

"Of course, but you're busy."

"Yeah. I suppose. If I get ahead of schedule maybe..."

"It would be great if we could go together, but I'll book a car anyway just in case. Should I bring my computer in here?"

She gave him her stupid question look, and he was back with it in a minute.

"Did you scan your transcripts and the MIT letter?"

"Yes – emailed them to you while you were asleep."

"Right."

He opened his computer and resent the attachments to Wilkinson. "About 2 am there now, so we won't hear from him for a while, I guess."

He scanned down the list of unread mails.

"Here's one from Prentice. Maybe the ABC has finally delivered..."

He opened it, and clicked on a link.

"Oh, *crap.*"

"What is it?" She jumped up and stood behind him.

He clicked on a link, and there was his interview with Schoor, on YouTube, where anyone could see it.

"The brazenness: Schoor has posted it himself. Look at his comment." His anger was something new to her – something she didn't want to see again.

```
A friendly person sent me this a few days
after it was made.  Truscott gained access
```

```
on false pretenses, unexpectedly aggressive
questions.  His movie will be more like
this.  Judge for yourself.
```

He jerked a finger at the screen. "The swine. He obviously thinks the best form of defense is attack. Put off others from talking to me; bias everyone against the show. Maybe even stop me from selling it."

"What else does Prentice say?"

Martin brought the mail to the front. "This is all it says:"

```
Brought to my attention by ABC management.
Any comment?
```

"So ..." Angie contemplated "Best form of defense is attack? What can we do to expose this for what it is?"

"Well, there's the question of where he got it. And why he's only put it up now."

"The second one is easy. He must have got wind that you were doing more interviews, maybe had a buyer."

"OK, so the first one ..." Martin contemplated for a moment. "The thing is, there aren't that many places it could have been leaked. My own computer is pretty secure. I encrypt the disk, I have a firewall, I close most of the ports. I have a honeypot, and nothing's been caught in it."

"Honeypot?"

"I have a fake login, with a ridiculously easy password. Once you are in, the standard command line utilities are there, but everything is rigged with a trip wire, so I will know as soon as someone tries to do something. Try it."

He opened up the machine's networks control panel and showed her its address. "Open up a terminal, and use `ssh` to log in – think of a possible login that could be there."

She sat down at her computer. "Like what?"

"Take a guess."

"OK." She typed in `guest`.

"Good so far. Now password?"

She tried hitting `return` for a blank password. Then got it second try – `guest` again.

"Hey, you're a good little cracker."

She nudged him in the ribs. For once, it hurt him more than it hurt her. "Oof. That was almost a cracked rib. 'Cracker' here refers to breaking security, not my bones. Now try a command."

She typed in `ls` to list files on her terminal session. As she hit `return`, his machine said "bing" and popped up a box with the message `strangers`.

"Neat. But how do you know that didn't happen when the machine was unattended?"

"This alert can only be seen by the real active user, and I log all logins to the bogus guest account. I'm pretty sure no one broke in."

"What about when you were editing, or after you gave it to Prentice?"

"First of all, the timing is wrong. He claimed he received it a few days after it was recorded – most people would mean less than a week, by that. Why he should say that I don't know, but that would place it before I left the US."

"Misdirection?"

"Maybe. But when we were editing with Julie, it was on

an even more locked down system than mine, and we did a really secure delete. Military grade. And Prentice's copy had a watermark on it."

"Watermark?"

"We added a transluscent logo on the bottom right. The small bird I used to talk to in the garden. Although this copy is downsampled, you'd still be able to see that – even if it was only a vague blur. And anyway, that copy had a slightly different edit of ... "

She smiled. "So who else could it have been?"

"NASA." His eyes narrowed. "The NASA people saw it. But they deleted it and emptied the trash while I watched."

"Like the secure delete at Julie's office?"

"No, just a regular emptying of the trash. Julie's took a lot longer. Someone with a file recovery program could have recovered it, if they acted in time. And that would have been only a few days after the interview – less than a week."

"So much for misdirection."

"Yes, but how to prove it?"

"Would one of the NASA people have done it?"

"I doubt it very much. They would be the last people to sabotage anything bad for Schoor. He is their mortal enemy – trashes their work in Senate hearings and the like."

"Could they have left a machine unattended?"

"Unlikely – they locked their offices behind them when I was there, and you need a password to wake the machine we used from sleep. I think a network breach is the most likely."

"Is this something we can prove?"

"I hope so. Let's see – it's ten here. What time will it be on the East Coast?"

"Plus three hours, 1 pm."

"If there's a network breach, can the phones be safe?" Martin contemplated. "Let's see. Blunt does Skype. We used Skype to show the movie to Wilson in England."

"Do you think that's how it leaked?"

"No, no chance of that. Even with YouTube's downsampling, it would look a lot worse than that off Skype. No, what I was thinking is that Skype is reasonably secure. The data stream is encrypted. If you have control of the local machine, of course, you can intercept the sound after it's decrypted. I think we should try to get him to use Skype away from the office – seeing as we have reason to be paranoid. Let's try this..."

A few minutes later, Angie made a connection to Blunt, with Martin off-screen.

"Hi Mart–"

"Yeah, I had an operation," Angie cut him off. "Listen, there's something our mutual friend needs looking up urgently in a Columbia University library. Is there one where you can look something up for him where you can use a university computer to Skype?"

"Well, I guess..."

"This is important. Do it. Please."

Blunt shook his head, then said, "OK, give me fifteen minutes. This had better be good."

A bit over fifteen minutes later, Blunt's ring sounded. This time, Martin took the call.

"OK, so what's the cloak and dagger stuff for?"

Martin used the chat feature to send him the address of the YouTube recording.

"What the hell?"

"I have a strong suspicion that someone recovered this from the trash on your computer. I went through every other possible source, and none is likely."

"How would they do that?"

"I don't know. Do you have a security expert who could look for evidence of cracking?"

"Yes. Our regular guy is pretty good."

"Not to be too paranoid or anything, but could you find someone independent, maybe from the university, to watch?"

"Yup. I think I know someone. Greg won't be offended.

"Why do you think it's more secure to use Skype here?"

"If your network is compromised, someone could have put a hack into your computer to record sound. Not terribly likely I suppose. Skype is encrypted so I feel reasonably safe with that, once it's on the network. Do you have an alternative?"

"My cell phone also does Skype through WiFi, so I'll call you back on that from the office. Maybe a bit safer than regular cell, if someone is watching us."

Martin said nothing about the possibility of bugs in the office. Presumably the NASA people would be cautious. Meanwhile ... cautious ... "Angie, I want to set a password on the house WiFi. Everyone will have to set it once on any computer before it'll work."

"Oh. Do you think someone's spying on us?"

"I doubt it but let's be safe. Do you have a manual for your WiFi box?"

They found the manual, and set up the password together, and checked that it worked on their computers. "We can set this up for dad and mom when they get back," Angie said.

A few minutes later, Skype announced a call from Blunt. This time, there was no video, but Blunt's voice was unmistakable.

"Got it! Most likely a techie contractor we had in the week before you visited. Just before your visit, he claimed he'd forgotten to install an update – insisted he must do it, felt bad about not completing the job.

"We found a key logger and he's created an open port on the firewall for himself, so he can log in on a nonstandard port without triggering the usual alarms on the firewall. Looks like he used the key logger to grab my password, then logged in periodically and did a search for movie files. Looks like he struck it lucky and saw it existed before it was trashed. Later came back and ran a deleted file scavenger."

"Pretty good. How did you track all that down?"

"For starters, he used his IP address in the firewall hole, so we know exactly where he's coming from, and he didn't think to delete the bash history file."

"Bash history file?" echoed Angie.

"A record of commands typed on the command line," Martin explained.

"I *know* that – I'm just thinking he must be an idiot not to think of that."

"Who's that?" asked Blunt. "Same as the person who called in the office? I thought it was you in drag at first, but the voice didn't quite fit."

"Sorry – I should introduce Angie, my partner in crime. And other things."

"Pleased to make your acquaintance. Anyway Greg is

going to close down the firewall port..."

"No – I have a better idea."

"What?"

"Tell me, does any of this constitute a criminal offence?"

"You betcha. Spying on a federal agency. A felony. We'll get the FBI onto this."

"But would this evidence stand up in court? Could you persuade a jury that you didn't create all those logs yourself?"

"Good point. I'll stop Greg. What's your better idea?"

"A honeypot. Stop Greg first, then I'll explain."

Blunt ended the call.

"OK, Angie, this is going to need careful timing. I wonder if Prentice can stir this up to a major news story. If we can catch these guys in the act of snooping on NASA, it will blow Schoor out of the water."

"Why don't you challenge Schoor to a debate on TV about your program?"

"I don't know. I've never done something like that."

"Come on. Armed with the facts, you can take him down. You did it before."

"OK. I'll see if he can swing it." The little boy talking to birds, the raccoon, the failed games programmer all faded into the past. He really felt he could do it.

He composed an email to Prentice.

```
Schoor stole this from NASA and I intend
to prove it.  Can you get media to run
a live confrontation between us?  I can
do it any time of day or night -- prime
time here, prime time Australia, to suit
```

```
Schoor's schedule, I don't care.  But soon,
while the evidence is hot.
```

"What's the time now in Brisbane? Eleven here ... must be something like 4 am there. Let's see what he can swing."

Blunt called back. Martin explained the honeypot thing, the plan to try to get a live confrontation.

"What if the FBI sting isn't timed right for your confrontation?" Blunt sounded a bit doubtful.

Martin had it all worked out. "It doesn't matter. It will get the media attention it needs so if the sting breaks later, it will still work. If earlier, he may refuse to show. Each way, we win. Just set it up, and get a search warrant organized.

"For now, I am taking some time out – I think I'll see if my swimming coach has anything else booked.

"Can I check back with you in a couple of hours?"

Blunt agreed – and Martin's swimming coach was available.

13   Showdown

A FTER THE SWIM, they headed home. As Angie opened the house, she said, "I don't know if I'll make an athlete of you. But you are getting better. Must be the good coach."

"I don't mind a naughty coach."

They walked back into the house, ready for whatever the world threw at them. It's amazing what a good swim can do to set you up for the day.

Martin checked the time – almost time for Blunt to call back. "Might as well check email in case Prentice is an early waker." There was nothing yet from Prentice, but a few local TV stations were asking if he could do interviews. He said he was not available and referred them all to Prentice.

"Good move," said Angie. "Show him there's interest."

Then Skype announced Blunt's call. "OK, it's all set. The FBI is watching the shop. We didn't tell the feds why we're timing it this way, but we've set up something really juicy to entice the perp into activity at around 8 pm to catch prime time. The rest is up to you."

"OK. Let me know when you're ready to go in. I still

don't know if I can get the media on my side – my guy in Brisbane should be awake soon. But either way, we've got him. I'll let you know developments from this side."

After Blunt disconnected, Angie mused, "8 pm eastern – that will be 5 pm here."

"10 am at home. So if Prentice can swing anything, . . . "

The phone rang. Angie picked it up. "It's for you."

It was Prentice, sounding totally shagged out.

"My damn phone has been ringing off the hook all morning, and it's only 8 am. Did you set the entire world's press onto me?"

"Sorry, but Schoor has blown this thing open, and we have to confront him – unless you have a better plan."

"Well, not now. Pretty lame of you to ask now. You might as well tell me what you're doing, since I have no choice."

"Not at all. You are the kingpin. I didn't realize giving your name out would have that effect. This is the plan. . . "

When he'd finished, Prentice said, "An FBI sting while you are on TV live? That should push up your shares."

"Yup. And yours. But we can't tell anyone about that in advance – spoil the surprise, especially for our sleazy friends."

"Right. So 8 pm eastern, you say. Not much time to organize, but we already have the interest. The local studios in your part of the world will be disappointed. Hmm. Maybe not. They should make a killing selling it out east, and they can rerun it for their own prime time. Especially if they don't have advance warning of the sting. I like.

"I knew it was a good bet putting my own money into this."

"But you said ..."

"Now kid, don't get all moralistic on me. I know a good horse when I see one. My accountant doesn't know which end of the horse is the ass, which is actually himself."

Martin sighed. "OK, let me know when I need to do something. Can I still refer all questions to you?"

"Yes. Knock yourself out. I'll get back to you in an hour, tops – you'll have to move if you aren't close to a studio."

"Berkeley is not too far out of San Francisco."

Prentice hung up.

"Awesome." Angie had never used the overused superlative before in his presence. "Sounds like it's all coming together."

"I hope so. I'm even getting used to people lying to me – hey, don't you get any ideas." She was looking mischievous.

"Who, me? You told me not to do that ever again."

"There was another 'that' you *were* going to do again." He did a passable imitation of her impish grin.

*   *   *   *   *

Half an hour later, Angie said, "Did you hear something outside?"

She walked over to the window.

"Oh, crap –" and ducked out of sight. "Did you see *Life of Brian*?"

"Yes, is it still popular here too?" he asked cheerfully, not taking in that something was happening.

"If you don't want to repeat the scene where Brian reveals himself to the world, stay away from windows."

He was instantly alert. "What do you mean?"

"The front yard is full of TV crews and press."

"Did they see you?"

"I hope not – but they didn't ring the doorbell, so maybe they think no one is home. From what I saw in a brief glimpse, they are all facing the street. Let's take a shower quickly and get presentable."

She herded him to the bathroom. "Together, to save time."

A few minutes later, with her fingers working his hair, he said, "I just realized two things."

"Oh?"

"There're a lot of good reasons besides saving time for sharing a shower and ... we didn't tell your parents ... "

His hair washing suddenly sped up. "Let's get out of here, and call them before they walk into this mess." She wrapped a towel around herself, then skidded to a halt, realizing the bedroom window was still uncovered.

He laughed. "Look, if they didn't see you first time, they probably can't see in."

She darted in, found some clothes, and brought them out. "You stay here, and I'll get your clothes. Maybe they saw me, maybe not. But you may not be as invisible."

As she pulled on her clothes, she somehow managed to call her mother.

"Grad student meeting? I know ... but this is *urgent* ... "

There was a pause as Melissa went out of the meeting. Having covered herself, Angie returned to her room and threw Martin's clothes at him. She blurted out the whole story, her mother interrupting a few times for explanations.

Her mother ended with "Stay right there. I'll drag Bob

out of whatever he's doing. If I can – otherwise I'll be right there, five minutes, tops."

The phone rang. It was Prentice. "Listen, the networks are keen."

"I'll say. Half of them are camped out on our lawn."

"Well, that's naughty, because the big event is happening in San Francisco – must be the ones who aren't in on the deal. No idea how they found out where you are. Tell them 'no comment' and barge through. You're on the air in an hour – can you make it?" He read out the address. Martin wrote it and showed it to Angela.

"Thirty minutes, if the traffic isn't too heavy."

He repeated this to Prentice. "Good. Can you get going right away?"

"In a couple of minutes. Angela's mum, maybe her dad too, is on her way. I think we need a bit of support to get through the crowd."

"Yeah, but don't be late – you have to do makeup first."

As Prentice dropped the phone, Martin said, "Time to catch up with Blunt."

He went to his room, and started a Skype connection. Blunt picked up instantly. "We're on air in just under an hour, eight eastern. How's it looking on your end?"

"Good. We can start rolling immediately – it should keep him busy for at least an hour."

"Great. Can I give you Angie's cell number? We will probably be OK using cell – there won't be time for someone to intercept and react – if they are at that level of surveillance..."

"Angie?"

"Partner." It was starting to feel natural saying it.

"Oh, right. The one who did the impersonation of you in drag. Let's have the number."

Martin read out the number, and wrote down Blunt's. "Gotta go – I hear a car showing up. We have to fight through a crowd of press."

"Yeah, you could hit one or two for me."

The car arriving outside was Melissa's. Bob was with her. As the press converged on them, Angie and Martin used the distraction to charge through the mob, and somehow managed to get in the car without taking the press with them. Cameras were flashing all around. Martin opened his window, and beckoned to them. As the mikes and cameras converged on him, clearing a path in front of the car, he smiled broadly at the cameras, then turned to Melissa and shouted, "Hit the road!"

As they were heading into San Francisco, Bob asked, "Who taught you to handle the media like that?"

"Huh. That's nothing compared with seeing off a sea of space aliens with super lasers."

"So all that computer gaming had some use after all?" Melissa smiled.

"Oh, crap!"

"What?" Melissa wondered if she'd said the wrong thing.

"That reminds me: I didn't tell my dad about all this. We haven't been so close and I just haven't thought about him."

Angie said, "From what you say, you seem to be kind of reconciling with him. I can't imagine he'd want to miss seeing his kid on prime time national TV – if they show it there."

"I don't have his number with me – maybe Prentice can help, and he'll also know if it's showing in Australia."

Prentice's number was busy. "Damn."

"What about Skype?"

"What about it, Angela? We need a WiFi hot spot for it to work."

"This is San Francisco. There's bound to be one just about everywhere. When we stop at the studio, I can call him, if you are in too much of a rush."

They were a bit early for the afternoon traffic, and made good time. It was just over fifteen minutes to broadcast when they got there. Martin typed his password in for Angela, and she found his dad's office phone number. He said, "One final bit of business ... can I keep your cell phone to keep up with Blunt?"

"Sure. Go inside. I'll talk to your dad."

He found himself being ushered into the depths of the building.

In makeup, a crew applied various layers with practised speed.

"There you go, kid." One of them said. Martin stood up, looking for the way to go next, then remembered Blunt.

"I need to make a phone call – can you duck out for a minute?"

The makeup crew grumbled a bit but left him alone. He called Blunt. "How's it going?"

"Worked like a charm. The connection is still live, and the feds are going in."

"So it wouldn't be inaccurate to say an arrest is happening as we speak?"

"Right. Good luck with the show. We'll all be watching here. I'm in a bar facing the crime scene, and they are announcing you."

"I don't need luck. This will be a slam dunk."

Martin looked for the crew. They were right outside. He didn't particularly care if they'd overheard. It was too late to do anything even if they had.

"We go this way. Ever been live on prime time before?" the talkative one asked, as they neared the studio. He affixed a wireless mike, as Martin looked in the door, noting Schoor on a big screen, already in the Boston studio.

"No."

"Could be scary. We can fix you up with a diaper if you need it."

"Fuck off," Martin snarled, striding to the door, adrenalin at an all-time high – previous resolutions on language forgotten. He turned back, realizing the guy had only been trying to lighten the moment. "FedEx it to Schoor. He's going to need it."

He stepped inside, and was directed to a chair facing several cameras. "This is it," he thought. His heart was going at a considerable rate, yet he felt completely calm and in control. "Let the facts speak for themselves."

The floor manager pointed at someone Martin hadn't been paying attention to, raising three fingers. He didn't know much about US TV. This must be the anchor. The fingers counted down.

On his cue, the anchor started talking. "Good evening. I'm Dan Gordon. In a special live event on Sixty Minutes, we are bringing you one of the top dissenters from climate orthodoxy, Damian Schoor, professor of physics at Harvard, with allegations against an Australian journalist, Martin Truscott.

"Professor, you posted an interview of you by Truscott on YouTube yesterday. What was that about?"

"This Truscott ambushed me. He claimed to be making a documentary giving fair coverage to the climate sceptics, and, as you can see, the interview was anything but fair. Hostile I would say. Extremely hostile."

"We'll play the interview in a minute. But let's see what Truscott has to say about this."

"I had an introduction from Doctor Kevin McCarthy, one of the leading Australian climate change doubters. I suspect Schoor –" he pronounced the name deliberately, making a point of leaving out the honorific, as had been done with his name "– did not pay attention to my email, where I said I wanted a full and frank discussion of the science."

"The science? You as good as accused me of bias!"

"You chose to publicize that clip. How do you know the context it would be used in, in the final edit? In any case, you couldn't answer fairly basic questions on the science. You put it out there. Let everyone judge."

Gordon intervened. "Well, professor, that is a point. Why didn't you answer the questions?"

"It was the aggressive tone, the supposition that this nobody knew enough to pick apart my peer-reviewed science..."

Gordon picked this up. "But professor, how many Senate enquiries have you fronted? You can handle aggressive questions."

Truscott spoke out of turn. "But *is* the Senate so aggressive? I've viewed a fair number of recordings and read a lot of transcripts. All the aggressive questioning seems to target scientists in the IPCC camp, not the contrary camp.

"I'll tell you what this is about," he went on quickly. "Reputation. You are trying to kill my reputation before this thing is screened, to put others off from talking to me, to bias audiences against the show."

"Reputation? What reputation? I am a Harvard professor. Who are you? ... "

"Gentlemen ... " Everyone was ignoring Gordon. Truscott wondered if he usually had such loose control of his shows. Surely not: he had to be one of the top anchors.

Truscott turned to the image of Schoor, and said icily, "Just because I am not famous, it doesn't mean I lack pride. My work is good, because I get the details right. Integrity means everything to me. Perhaps, on that note, you can tell us where you got that clip."

"I have friendly sources. You think climate scientists all follow the party line. The silent majority ... "

Before Gordon could take control, Martin calmly smiled. "Tell me professor, do you think spying on a federal agency is a felony?"

"What do you mean?"

"That clip was sourced from a security breach at NASA. And right now, as we speak, an arrest is being made."

Martin wished the makeup didn't obscure the fact that Schoor had gone white. He had to imagine it. Then the reason for Gordon's inattention became clear.

"An arrest being made? I can confirm in breaking news from New York, premises of a computer company are being raided. Let's go to New York..."

Schoor was ripping off his wireless mike, and making an undignified exit, as another screen in the studio switched to the scene in New York of several computers being carted into an FBI truck, and a disheveled handcuffed person being bundled into an FBI car.

The rest of the segment was a blur. Martin hoped he had said sensible things about the NASA break-in. In New York, Gregory Margolius was introduced as a NASA security expert. He said something about how the arrest was made. The YouTube clip was shown. Martin was asked questions about the rest of the show, which he deflected on auto pilot.

As he left the studio, the makeup crew was waiting. The talkative one was brandishing a FedEx box.

"What's this for?"

"We were just wondering if you had Schoor's address handy."

They led him to a waiting area, where the Greens were all watching a big screen. Gordon was wrapping up before moving on.

"That was ... incredible." Angie nearly knocked him off his feet, in a passable imitation of their meeting at the airport. After a moment, she added, "Your dad said he was going out to a bar to watch. Seems one of the commercial channels over there ran it live." She looked thoughtful, and held him at arm's length. "Why did you suddenly remember your dad, when you did?"

He laughed. "He was constantly slagging me off for wasting so much time on computer games. When Melissa said how all my computer gaming had some use, I instantly thought of him.

"You know, he was actually right – in general terms. I don't know of any other life skill but seeing off a bunch of hungry press that computer gaming could apply to. Slaying orcs, trolls, space aliens . . . " he stopped, realizing some of the TV people were still there. But they joined in the laughter.

Bob took charge. "Let's go home. It will all be repeated here at eight. You want to see yourself perform."

Martin added, "And it's my turn to cook. I really want to," he added, to staunch protests. "Let's go shopping on the way home. I want to get pizza stuff, and a nice chardonnay to celebrate victory over the dark side . . . they do get chardonnay here, don't they?"

"Of course, dear," Melissa said. "But only at liberal liquor stores." The TV people got it too. They were led out to their car, in a festive mood.

As they drove to the store, Angie leaned forward to her parents. "Mom, dad. A little announcement."

They paid attention to her, Martin wondering where this was going. This sounded formal.

"Martin started calling me his 'partner'. Do you mind? I kind of like it." she asked, unusually coyly for her.

Melissa glanced back briefly, her focus quickly back on the road. "Of course not, dear."

Bob nodded. "Two celebrations in one day. I'll pick out a really good chardonnay."

*   *   *   *   *

Pizzas were emerging from the oven by the time the TV show started.

"That's one thing I forgot – without a commercial-size oven, you can't do a lot at once."

"Shut up, Marty." Angie dragged him onto the couch. "Forget cooking – let's watch.

"Great pizza, though." Everyone nodded.

"Nice chardonnay too," Martin added ... then he was on.

He sat through it in something of a state of shock. Had he really pulled it off? Events of the last three hours smothered him – the adrenalin finally wearing off. The congratulatory calls from his dad, from Blunt, then the flood of press calls, all referred to Prentice, until Bob unplugged the phone ... He collapsed onto Angie like a sack of potatoes.

He had a vague recollection of being dragged to a bedroom by more hands than he could count, and an even vaguer recollection of being undressed by fewer hands, and being put into bed.

14   Partner

ARTIN WOKE UP, a feeling of a huge weight having lifted. He turned over, and Angie was there. He looked around. This was her room. He touched her lightly, and she snuggled up to him. What exactly had happened last night? What the hell. He reciprocated the snuggle and dozed off again.

It must have been at least an hour later when he awoke again, this time with sunlight sneaking past the curtains. He turned clumsily, and there she was again, this time awake – mainly because his elbow caught her in the ribs.

"Hey, mister. Watch those elbows."

"What happened last night?"

"You passed out."

"But I mean how did I get here?"

"You levitated – a better trick than sleep walking."

"No, but I mean, we weren't sleeping together ..."

"We're officially partners now, aren't we? Or does that mean something different in Australia?"

He showed her what it meant in Australia.

*   *   *   *   *

It was almost nine by the time he looked at his watch. "I suppose I'd better get up – check my appointments. I think I have some set up ... things are a bit hazy after yesterday. Meant to be my rest day ... "

"You stay right there. I'll bring you your computer."

She was back in a minute.

"Wow. Room service."

"Yes," she said. "Good point. I'll get some coffees together while you check your mail."

"What about you – your classes?"

"Screw that – I can afford an A-minus for once."

Only an A-minus? He raised his eyebrows. So competitive. I guess I need to get that way if I do the PhD thing. Then again, Schoor may think I'm pretty competitive. He grinned, trying for that Angie naughty look, and found a mirror. Got to get it down before I unleash it on Ange.

He scanned down the email headers. Three, no four... five from Prentice. He reordered the headers by sender. No, those were all the mails from Prentice. Let him wait – he'll be asleep now anyway.

A couple from Wilkinson. He took a look, and left them open for Ange.

And Blunt. Short and to the point:

    www.nytimes.com

He clicked the link. The lead story had a picture of the FBI raid, with his name in the first paragraph. He started to read, as Angie opened the door, with a pair of cups, and a paper under her arm.

"Latte?"

"Too early for chardonnay?"

"If it puts you to sleep like that . . . look. Mom or dad must have got this." She put the cups down and tossed the paper on the bed. It was the *New York Times*. He pointed at the screen. "Oh. Beat me to it."

"The paper edition has a bigger picture. I'll read it there.

"You take this, and look at these." He handed her the computer, with Wilkinson's mails brought to the front. "You have options at UNSW and Sydney Uni."

"Sydney Uni? How good is that?"

"Also one of the best. It comes down to what you want – who you'd like to work with. One of these or. . . MIT."

"We haven't heard back from MIT yet." She sipped the latte.

He tried some too. "Good latte.

"OK, we were going to start that whole thing again. Tell me what you *really* want."

"MIT is not an option unless. . . "

"Look at this – I'm all over the front page of the *New York Times*."

He took the computer back and scanned over some of Prentice's mails, without taking in all the detail.

"I could be booked solid until next year on TV shows if I wanted. I'm a celebrity. I really feel now that I can do anything I want. It matters to me that you do what you really want. I can fit in."

She took a deep breath.

"MIT was only ever a way of getting some time away from home. I've always wanted to go to Australia. I'll check these out and if any of them looks good, I will be really happy to go there."

"But you will definitely check them out and *really* make sure you end up doing something you want?"

"Drink your latte. It's getting cold."

"I take that as a yes."

"Use your interview magic on me to make sure I'm not lying."

He gave her nose an affectionate tweak. "Is anyone else home?"

"No."

"I checked my appointments – nothing at Berkeley for a couple of hours. I'm seeing David Wu about predictions of increasingly violent storms. Then a couple of research groups to verify facts, no filming aside from Wu. Prentice can wait – he'll be in bed now.

"Should we explore the shared shower thing again?"

"Then go for a nice wake-up swim?"

*   *   *   *   *

Martin left her at the pool. The swim felt better than last time, though he was still nowhere near to her class. He thought he glimpsed some of the gang heading in, on his way out, but checked his watch... no time to stop and talk. It was a strange experience, striding to David Wu's office. It seemed as if he had last looked up a Berkeley professor years ago – as a different person.

Wu was waiting anxiously for him.

"Great you could still make it! I was worried that with all the media attention, you'd be sidetracked."

"I'm here for one purpose. The other thing was a distraction."

"Yes, yes. Of course. So pleased to meet you."

Wu looked nervous. Martin laughed. "It seems I have some reputation now for chewing up and swallowing people whole. You have no idea what Schoor did to provoke all that."

Wu brightened up a bit. "Yes, yes. You know, I went to one of those Senate hearings, and it was exactly as you described. They took me apart as if it was a court of law. No regard for arriving at the truth. No regard."

Martin belatedly shook Wu's hand. "I'm so sorry, I have only recently upgraded from being a computer geek. My manners are terrible."

Wu relaxed visibly, but still looked tense. "Professor, where would you like to conduct the interview? I prefer natural light, but it's a bit blustery out. Is there a nice space in a building nearby where I won't need extra lighting?"

The interview focused on the prediction of worse storms as the climate changed – how warmer water causes stronger tropical storms, worse cyclones and hurricanes.

Wu was so nervous, that Martin stopped after a few minutes.

"I'm sorry. This isn't going well. The person you saw on television isn't my usual persona. Can I get you a coffee or something, and we can just have a chat?" Wu agreed.

After five minutes of talking over coffee, Wu forgot his nervousness, and discussed most of the issues Martin had hoped to cover. Kind of like the small boy trying to win the trust of curious little birds. . .

"Can I set up another time for the interview? I hope you

can see now that I'm not such an ogre."

Wu nodded appreciatively. "Let's go back to the office and talk to my secretary. I'm sorry about this whole episode. I've never had this happen before."

"Believe me, professor – this is new for me too. And not a welcome development. That whole thing was more a reflection on Schoor's personality than mine."

The next two appointments with the research groups covered ground he'd already been over, and there were no big surprises. The grad students were in awe of him, and their professors were nervous, but, without a camera in sight, it went off OK.

At the end of it all, he was glad to find Angie and go home.

*   *   *   *   *

That night, back at the house, Martin told the others about his experience with Wu.

Bob looked thoughtful. "All you need to do is carry on as before – cut people a little slack if they are nervous. Before, no one – except that idiot Schoor – thought anything of talking to you. No pressure. Now you have a fearsome reputation. You cut a Harvard professor to shreds on prime time TV. It frightens me, and I know you're a nice kid."

Melissa asked, "What of Prentice? If he's lined up some media time for you, you could soften your reputation. If everyone is scared of you, Schoor has won – at least the part of the battle where he gets to stop you finishing the movie."

"Oh, crap." Martin ran upstairs, shouting down to them as

he ran. "We didn't check Prentice's mail in detail this morning, because it was night time over there."

"As long as I don't always have that effect on him." Melissa smiled.

Her daughter grabbed her in a bear hug. "No, mom. You just have a way of saying the right thing at the right time."

A few minutes later, Martin headed downstairs, carrying his computer.

"Just as well we remembered." Angela pointed at Melissa, and he corrected: "OK, just as well *Melissa* remembered.

"Prentice has set up some interviews for me, and some time on a radio talk show... Uh oh."

He went to the nearest phone, and plugged it in. It rang straight away. He picked it up. Sure enough, it was Prentice, cursing a blue streak.

When he stopped, Martin brought the phone closer to his face. "Look – I'm really, really sorry. But the phone wouldn't stop ringing..."

"Never mind. Did you read your email?"

"Yes – I have a talk show tonight."

"Right. You don't have to go to the studio if you don't want, but it will work a whole lot better if you have a working phone. They have been trying to call you for the best part of a day."

"OK, and the TV thing tomorrow..."

"Several TV things, if you can fit them into your schedule..."

"Actually, yes. I want to do them before I do much else." He told Prentice about Wu.

"Yeah, I'm a bit scared of you too. Is it true that you bit one of the makeup people in half?"

"Almost. But that wouldn't have been very vegetarian. I'll do everything you've set up so far. I hope it repairs the damage."

"Good boy. By the way, your dad called me – asked if you've set up a company."

"A what?"

"Kid, the way this thing is going so far, most of what I'm paying you, you can write off as tax deductions – all that travel, subsistence per diems... but when the main money starts coming in, you want a company to keep the government's grubby paws off the money."

"So, what did you tell my dad?"

"Not yet. So he says to tell you that he'll help if you want."

"Oh."

"Is it a surprise that your dad wants to help you?"

"No. it just occurred to me that with the phone out, he has no way to get in touch. We've never talked by email."

As he put the phone down, he said, "I'm such a rat. This is the second time in twenty-four hours that I forgot I had a dad."

Angie went up to him, and carefully examined him from top to toe, with a bit of judicious pinching, prodding and pummeling.

"Dad, mom." She turned to them.

"You didn't spend all that good money to keep me in the top tier of the UC system for four years, just so I didn't know my biology."

She turned back to him, put a hand on each shoulder and looked him straight in the eyes. "Almost certainly one of the great apes. Probably a hominid. But *rattus rattus*? No way. My partner has to be at least a great ape."

Once they had recovered, Angie looked thoughtful. "Well, there's the phone. Talk to him before other duty calls. I'll make sure you don't forget anything important." She handed him a portable handset.

He went upstairs to find the phone number, and walked downstairs talking on the handset.

"You saw it? Great...

"What do you mean, who put me in charge of the FBI?...

"Is that what it looked like? No wonder everyone is scared of me. I would be scared of me...

"What's this story of Prentice's about setting up a company?...

"I know that you've been running your own business for years... yes... thanks, that would be great."

He was about to end the call, when Angie grabbed the phone.

"Mr Truscott," she began tentatively...

"Is that Angela?"

Martin overheard this part and said, "Yeehaa! He remembered her name this time!"

Angela put her hand over the phone, and said, "Shut up and listen to what you forgot – aside from that you have a dad."

He looked suitably chastised.

"Yes. I thought it was about time your son's partner introduced herself."

There was a silence at the other end. "I'm sorry, I'm a bit out of touch – does this mean something like that the two of you are married?"

"Not exactly, just that we do everything together."

"Well, if the two of you do get married, don't forget I still exist."

Martin grabbed the phone back. "Dad, things have been hectic. I am so sorry I haven't told you everything. Give me your email address and I promise to do better."

He wrote it down.

They had a low-key meal that night. Bob and Melissa took care of the details.

The talk show was a new experience for Martin, who didn't listen to talk radio much.

The first question phoned in was predictable: "How did you get the FBI to do that?"

"What?"

"Time their raid for prime-time TV?"

Martin laughed. "Schoor was right about one thing. I really am a nobody. No one would have heard of me if he hadn't behaved the way he did. I didn't put that clip on YouTube. He did. So, I guess you almost by definition have to be more important than me. Can *you* think of a way you could get the FBI to jump when you want them to jump?"

"No, but... you *did*, so how did you do it?"

"Looking for hints? The simple answer is, I didn't. The NASA people went looking for a security leak because they'd seen the clip. Schoor was good enough to take up my challenge to a live TV confrontation. Ask him how *he* set it up."

There were questions about climate change, all the old

claims about bogus science etc.   Martin answered some straight. The fourth time he said, "Look at it this way. A whole bunch of scientists have been working on a major, major problem for decades. Someone pops up who has done no work in the field, and says, the scientists are a bunch of doofuses – they're missing something obvious. So what do you do? Believe this and stop paying attention to the scientists, or go and ask the scientists if it's true?

"That's where I started from. And I chose to ask the scientists. That doesn't mean I believed everything I was told. I read at least a dozen papers before every interview, and have read over a hundred now. I've also spoken to the major players on the other side, and worked hard to understand their work.

"You want to find a hole in the science? *Study* it – don't believe someone who has an axe to grind. They may be right, but very, very few scientific dissenters turn out to be another Galileo."

At the end of two hours of this he was tired, but elated. The first hurdle. And the word "partner" had an increasingly special feel to it...

As they climbed into bed that night, Angie said, "The gang missed you at the pool today. Dale says he has something he has to tell you – wouldn't say anything to me. They were all over me to tell them how you pulled off that stunt, but I think you should tell them yourself."

"OK, so swim tomorrow... I think you need to get back into classes so how about I tag along, and take some time the first half of the day to work through which media to see when? I'm sure no one will notice me if I sit in a lecture doing it."

"Yeah, we'll bring a big towel this time."

"Huh?"

"In case you have another attack of modesty."

He grinned, remembering the incident, then looked thoughtful. "When you did that *rattus rattus* thing, it was just as well I didn't tell your folks I used to be a raccoon."

She carefully inspected him again, leaving out no detail. "A wonder of biology. A hominid with a raccoon juvenile form. I'll have to study you for my PhD."

Martin looked contemplative. Then: "As long as you also do something you can write up as a thesis."

15 Celebrity

7:30 AM. There was a phone ringing. Martin blinked sleep out of his eyes, and Angie fumbled for the phone near the bed, jamming an elbow into his solar plexus.

"Oh no, let me kiss it better. I really didn't mean to retaliate for yesterday."

While he was recovering, there was a tap on the door. It was Melissa. "Martin, it's for you...can you pick up the phone...the *Chronicle* wants an interview...if you want to talk to them..."

"Just a minute!" Martin looked for his list of media. "OK, I'll take it."

He picked up the phone.

"Sorry the phone was busy all night – talk show. I did mean to get to you first thing."

"Yeah. Your agent was pretty insistent that we didn't harass you – no call later than 10 pm. José Mendez. We're shooting for a big story tomorrow, since it's not breaking news anymore. Background, where it all started, and so on."

"OK."

"Would you mind taking some questions now, and can we

do a photo shoot later?"

Martin explained his schedule for the day.

"Around midday on the Berkeley campus will be perfect. Do you have a location we can meet or a cell phone?"

Martin conferred with Angie, then: "As long as you promise not to give it out, here's a cell phone number." He repeated her number. "We'll expect your call around twelve."

Mendez asked a series of questions, some covering similar ground to the talk show. Martin wondered if he'd listened in, but decided not to quiz him on technique. He explained the NASA connection, how he'd shown the NASA people the clip, then watched while they deleted it.

"So you didn't think anyone would know you showed it to the NASA people?"

"I didn't really give it a thought. I wasn't sure if I should have shown it to them, and made them promise not to tell anyone."

"How did you come to suspect a NASA break-in?"

"I knew my own computer is secure, because no one else touches it, and I have a few tricks to detect intruders. Schoor said he had the clip within days of its being made. NASA was the only likely possibility. I alerted NASA, and their security people did the rest."

"Including timing an FBI raid to coincide with your prime-time interview?"

Martin shrugged. "The perpetrators didn't have to break into NASA at exactly the right time."

"You realize some people are saying this is a set-up, you did this to score maximum publicity."

"I'm sorry, but I didn't put that clip on YouTube. I have

from the start gone after the science. Schoor went after me – he was trying to wreck my chances at being seen as objective. What was I supposed to do?"

"I don't know what you were supposed to do, but what you did was remarkably effective.

"That's it for now. Anything you want to add, off the record?"

"What does that mean?"

"I won't report it unless I can do it in a way that doesn't source it from you."

"No, not really – if I have anything to say, it will have my name on it."

"Good talking to you – see you later."

He hadn't noticed Angie sneaking out, but she timed her return exactly – "Room service again!"

"Before you totally spoil me, I'll have to do this for you tomorrow."

She brought in a tray with muesli, lattes, orange juice and melon. "Thought the cantaloupe looked good..."

He sniffed it. "Smells good too. We have a different name for this kind of melon in Australia – can't remember what..."

They went over the schedule again over breakfast.

"OK, so you start with computer architecture. I'd like to see what that's like here, maybe I should focus on my schedule after that."

They discussed other activities, and decided to meet at the pool again at midday. She handed him her phone. "Better take this now so I don't forget to hand it over."

They arrived in the class. It wasn't as big as Martin expected. "Isn't this a popular course?" he hissed.

"Yes," she whispered back. "But it's a grad class. Did I forget to mention?"

He sat near the back with her, feeling more conspicuous than he wanted.

The professor strode in, and started making announcements about a project, then saw Martin, and stopped dead.

"You're that guy who shredded that asshole from Harvard, aren't you?"

Martin suddenly lost his swimming tan. "Uh, ... "

"Come here – it's not every day we have a celebrity in the class. Show yourself off."

Martin walked down to the front, feeling totally out of his depth, the class cheering as he walked.

"Uh, guys, I'm just an unemployed CS graduate making documentaries until something better shows up. I think the rock star who looks like me is next door."

The professor shook his hand. "If any of my grads has the impact you've had, ... " He looked at the class. "We always tell you about industrial impact, doing stuff people can actually use. This isn't quite the same as designing the next big thing in microprocessors, but it's still pretty big." He looked back at Martin. "Welcome to Berkeley. Pleased to have you here."

At the end of the lecture, Angie shooed off classmates. "Guys, my man has work to do. So do we. We meet at the pool at 12:45, OK? Not a minute earlier."

As they walked out, Martin asked, "12:45?"

"You have your photo shoot, and Dale wants to talk to you, remember."

At about 11:45, Angie's phone rang. Martin pulled it out

of his pocket. He explained where he was – heading towards the pool.

"Perfect, there's a place right there were we can get some good shots. I'll be there in five."

Mendez brought a photographer, who snapped at Martin as Mendez asked more questions, then positioned Martin in various poses. Martin baulked at saying much about Angie. "You can say she's my partner, OK – but anything more, and I don't talk to you again. She'll be a bigger celebrity than me in her own good time."

"Great. I think we may still make page one. Local angle."

"Just one thing . . . "

"Yes?"

"Any chance I could get any of those photos? This spot has good memories."

He saw off Mendez and the photographer, then went into the pool facility to change. Angie was waiting with Dale for him as he emerged. "Ready for action, I see. You two talk while I change. How did the *Chronicle* go?"

"OK," he called after her.

Dale led him to a quiet spot.

"I guess you may be wondering why the press were all looking the wrong way at Angie's house the other day?"

"Now you mention it . . ." He told Dale of the Monty Python moment.

Dale had a fit of laughing. He punched Martin on the arm. "You guys owe me big time. My dad works for Channel 7, and as soon as I saw that YouTube thing, I knew you'd be besieged, so I called my dad, and asked him how we could get the press off your backs."

"He helped with that?"

"Angie is a good friend. If he hadn't helped, he'd have just been another one in the pack. So he had me call in fake sightings of the two of you. We had half the press searching for you over the whole bay area, and the other half gave up the chase and waited for you to show up at home."

"So we owe you. What payment would you like?"

"Well, if you could give my dad an exclusive interview, it would make him very happy, but that's up to you. For me, I'd just like a photo shoot of my new heroes."

Angie appeared as he said that. "A photo shoot?"

"Just a couple of pictures. For me, personal. I promise not to let copies out."

Angie looked doubtful. Martin quickly explained why they should oblige.

"OK, but I'm shy," she said, with mock coyness.

"Towel?"

"No. That only works to cover modesty. I'll just hide myself behind you."

She arranged herself behind Martin, not much more than her face pointing at the camera. "Anyway," she added as Dale took aim, "he's gay."

"Wonderful," said Dale, evidently not hearing the cause – "that stunned expression is perfect. One more?"

After a brief swim, they joined the now much enlarged gang. They were all over Martin, a barrage of questions worse than the press pack's. Angie took charge. "OK guys, we only have a few minutes before class. Dale was here first, so he gets first question." To grumblings of no fair and the like, she added, "Who makes the rules here anyway?"

Dale grinned. "We know who's in charge. Martin, what was it like going live on prime-time like that?"

"I was expecting to be really scared, but it seems to me I frightened everyone else." He told them about the offer of diapers, and how that ended up.

Pete jumped up to be noticed. "See? Told you you're a god."

"Yup, but I manage access. Pete, you had a question." Angie pointed at him.

"Yeah. How *did* you make the FBI perform on cue?"

"I'll never get rid of that – the FBI raided when a new security break-in started. Why those clowns chose to do it then is their problem. They knew Schoor was live on TV."

"But they didn't know they'd be *caught*." Peter finished.

"Yeah. Good advice to all slime bags. Always assume you'll be caught," someone added.

Angie announced, "Time up, we gotta go."

A voice from the back said, "High priestess..."

"No," said Martin firmly. "Chief goddess."

As the others left, Angie said, "I didn't know you were religious."

"Neither was L Ron Hubbard, and he invented a religion as a tax break." He grinned. "Anyway my personal belief system doesn't extend beyond whatever it takes to stay in your cult."

She wrapped a towel around herself with contrived deliberation, mimicking his maneuver of the first time they met the gang at the pool. "Uh, oh. I need to preserve my modesty too."

They laughed, then she asked, "How's your schedule panning out?"

"I need to retrieve my computer from the locker to check, but I'm pretty sure Seven was on my list."

They met outside the locker room after changing and he confirmed his suspicion. "Prentice gave me a few local channels to choose from, and if it'll make Dale's dad happy, I can strike the others off. Then, let's see... Al Jazeera, a Chinese network... Prentice has set up SBS – Australian channel... I wonder why not the ABC – Australian ABC?"

"Sounds like you'll be busy. Gotta rush. Here at five?" They were outside the pool complex. "Oh, and keep my phone. You'll need it."

The interviews went over pretty much the standard ground. Al Jazeera was the biggest unknown, but they turned out to be very professional and asked informed questions. The Chinese crew spoke good English. He guessed there'd be a voiceover in the Chinese broadcast. Anyway nothing would get through that the Chinese government didn't approve of. He steered the conversation firmly to the consequences of a major melt. "Half a billion homes lost in a five metre sea level rise... The question is not: 'Can China risk losing one or two percent of GDP to stop climate change?' The question is, 'Can China afford to lose Shanghai?'"

The SBS interview was pretty standard as well, but some local angle was obviously called for.

"How is your family taking your sudden catapult to the limelight?"

"My dad is taking it very well. We were never a close family, before but that's changing in a big way."

"Brought together by crisis?"

"Not really. I don't see it as a crisis. That's just the way the timing worked out."

"Where do you expect to go from here?"

"Finish the documentary."

"That's it?"

"No. But that's all I have planned out. Other things will follow, you can be sure."

"Martin Truscott, thank you."

The Seven interview was a little different because of the personal connection through Dale. Murray Brown introduced himself. "I really appreciate the exclusive. This will go to all the networks across the country, and some international as well. The ABC in Australia wants it, and some of the commercial channels there too."

Martin raised his eyebrows. Why is the ABC out of the loop?

"Dale is a good friend of Angie's and –" realizing he shouldn't disclose the deal in front of the crew, he paused, "I guess I feel a bit more comfortable with talking about the personal side with someone she knows. What I mean is, you know the limits."

"I understand. She's a great kid, and I know just what my media buddies can be like. If you don't want to answer a question, we'll cut it."

The interview followed the usual line, but Brown asked a few delicately positioned questions about Angie – how they'd met, future plans.

Martin answered openly, but shied away from detail.

At the end, he said, as the crew was packing up, Martin asked, "Good enough?"

"What do you mean?"

"Enough 'personal interest' to get that out of our lives?"

"There's never enough." Brown shook his head. "If you want my advice, ride this thing until you get your movie out, then lay low. Some of these people are vultures – even though I work with them every day, I get sick to the stomach sometimes when I see what they do to get a story."

Martin shook his hand. "Thanks, thanks very much, Mr Brown."

"Murray, to my friends."

"Murray."

That night, Martin and Angie were watching TV, surrounded by papers, with Melissa taking her turn in the kitchen, when Bob arrived. "You know, I think the worst of it is over. I had a call today from Wu. He's been watching the news, and is happy now that you aren't totally feral. Seemed to help that he's connected you now with a fellow Berkeley academic."

"Great. I need to reschedule all my interviews. . . see who I can still get. . . check again with the ones I didn't pin down." He looked thoughtful. "I think I'll ask Prentice what's going on with the ABC. They're supposed to be buying my movie so I would have thought they'd be first to do an interview."

Bob nodded. "You're right. It doesn't make sense."

Martin went up to the bedroom to make the call. Prentice's office answered on the second ring, and Martin was put straight through. "Bill," he said, "the ABC. I thought the deal was all but sewn up, just some signatures or the like. How come they aren't at the top of the list for interviews?"

"Because I didn't put them there. Seems like political pressure from Canberra, coal lobby or whatever. Bloody ABC used to be independent, bought stuff on merit. Luckily you're a hero now – Brisbane boy takes down Harvard prof – appeals to the tall poppy instinct."

Martin groaned.

"Never you mind, lad. Now they are falling over themselves to get access."

"So have they signed?"

"Yup. But only after I put them at the end of the list and threatened to keep them there..."

Martin smiled, but Prentice went on, not seeing him – "You should have heard them pleading when I told them I would drop them cold if they didn't sign..."

"You didn't!"

"Absolutely I did – and took international distribution rights out of the deal. This is going to cost them a bundle, and we're going to pocket the difference. I already have options from channels in the UK, France, China and South Africa. I told them I had my address book open, and it was just a matter of whether I called SBS, Ten, Seven or Nine first. My dialing finger was itching...so can you do a live interview on *Lateline*?"

"What time?"

"10:30 – you'd need to be in a studio half an hour before."

Martin groaned again, this time with real feeling. "You mean 5 am our time. I would have to get up at four to get there in time."

"Good. I thought you'd say no."

"I didn't..."

"Yes, you did. We'll fix up a recorded interview at your convenience, OK? Mail me some times you can make. And how are you getting on with the things I set up for you?"

Martin reported on the various interviews and the talk show the night before.

"Good boy – you did a lot. I like the ABC being forced to buy in an interview. I'll propose they use the interview from your local Seven for *Lateline*, and ask some supplementary questions – maybe even by phone. Yes, we'll get them to do that. Save you a trip to a studio."

Martin went down as Melissa emerged from the kitchen with some snacks, and shared the news. Angie pulled him down onto the couch. "Now you have no excuse for modesty."

"Oh. No more towels." They had to explain.

Melissa was heading back to the kitchen when the phone rang. Bob took the call. He brought a cordless handset over to Martin. "It's for you. Lovecott at Stanford."

Martin took the phone to a quiet corner. He received a blasting. When Lovecott subsided, he brought the phone back to his ear. "I didn't set up an interview with you because you never replied to any of my emails... of course I want to see you... at the top of my list... when? I'm rescheduling everything so I'm sure I can work to your plans. I'd love to visit Stanford over next weekend – can I just check?... Call me back in five minutes."

He walked back to the others. "Wow. I didn't think I had much chance with him – he's had major run-ins with the press over the years."

Bob offered Martin a book with Lovecott's name on the

cover. "Will you have time to read this before you see him?"

Martin took the book. "I'll try. He wants to see me on the weekend. I'm thinking Sunday, then I can try to swing a few more interviews on Monday, if I can find a place to stay overnight. If no one has anything planned, how does a family trip to Palo Alto sound?"

They all agreed it sounded good.

When Lovecott called back, Martin picked up the phone. He explained the family trip idea – "It's OK, they're all nice." He explained who Angela was, how her mother was a Berkeley physics professor, her dad an economics professor. "What do you mean the bad kind of economist?. . . Oh, you mean like the clowns who write in the *Wall Street Journal*?. . . I sincerely hope not. . . anyway we can both work on him if it comes to the worst. . . Good, we'll bring a picnic."

"What was that about?" Bob demanded.

"Economists who think there's a choice: environment or economy."

"I see. I have read his book, you know. I kind of think we'll get on."

16 Stanford

S TANFORD. One of the great universities of the world.
Properly known as Leland Stanford Junior University –
founded in the nineteenth century when railroads were
the basis of fortunes, by railroad magnate Leland Stanford –
it was named for the founder's son who died at an early age.

The campus, with its mix of Spanish-inspired sandstone
and modern steel and glass, is one of the most pleasant academic sites in the world. With hills in the distance, and no
great city to impose its stamp, Stanford dominates its locale,
without being separate from the world. That also captures
its tradition: fearless scholarship rooted in reality. Whether
in biotech, computer science or business practice, Stanford
has had massive impacts – the invention of many concepts
that drive modern industry, masses of publications in top-tier
journals and conferences.

Yet the thing which really makes the place stand out is its
diversity. What other great university has both driven a high-
tech revolution, and pushed the bounds of what is sensible to
do, ethically?

James Lovecott was at the centre of this maelstrom – a

bioethicist who had nonetheless made major contributions to industrial best practice.

He met the Greens and Martin in the quad, an area surrounded by the older-style stone buildings. "I thought it would be nice to take a walk up into the hills. Some nice clear air, a site or two where I can make some points, and of course, we can picnic up there as well."

They all followed his energetic strides.

"Come on. We can walk for hours without exiting Stanford land. But I want a space where we can get some perspective."

Up in the hills, they paused to catch their breath. Lovecott pointed back the way they had come. "It all looks so small and insignificant once you get some perspective. Let's go higher."

They followed, Lovecott pointing out landmarks. "There's Stanford Linear Accelerator Center – SLAC."

Eventually he found a spot to his satisfaction.

"Do you see any butterflies?"

They all looked around.

"Of course not. It's the wrong time of year. Late February to early May, you'd see hundreds of checkerspot butterflies right here, then they'd lay eggs and die. Beautiful."

"Pity we aren't here earlier in the year," said Melissa.

The others looked at her disapprovingly.

"Ah. I see one person didn't read my book. They all disappeared from here twenty years ago."

"But not because of climate change," Martin added.

"Right. Let's do a quiz. Anyone remember the reason?"

Bob put up his hand. "Ah. An economist who knows about butterflies. Bob, what's your answer."

"Invasive plants."

"Exactly. Seeds from gardens, hay brought in for horses."

He stooped and pointed at a plant with thick pointy leaves. "Dwarf plaintain. Not many of these left here – out-competed by alien plants. The larvae used to feed on these... it feels like lunch time." This neat segue dispatched, he broke out his lunch, and the Greens unpacked their picnic.

As they worked through sandwiches, Melissa said, "You'll have to excuse me for being an ignorant physicist, but what does all this have to do with Martin's movie?"

"Aha! Ignorance is not so terrible if you know you're ignorant and ask the right question."

Lovecott waved a sandwich to emphasise the point, and went on. "Everything is connected. Those butterflies, these plants. Change something, and you throw the system out of kilter. Maybe you're lucky, it adjusts, nothing much changes. But look here at what a ridiculously small change has caused. Some unintended grass, weed, domestic flower seeds. Whatever. And we've lost this butterfly habitat for good.

"Yes, for good. We did an experiment of trying to get the habitat back – but it didn't work. The intricate relationships are broken. It's not just the one plant, the one animal. Everything's linked.

"Now perturb the system in a big way –" Lovecott's monologue continued – "move up global temperatures by one or two degrees, let alone three or four. What do you think will happen?"

"I really don't know," said Melissa.

"Exactly. None of us knows. There've been big changes in the past, but we weren't there to witness them. The fossil record is incomplete. Very few invertebrates, relatively."

"Amber," added Angie.

"Exactly. Preserved in tree sap that set like rock. How often does that happen?

"But now we are conducting this huge experiment. Not just on us, on the entire planet." Lovecott expanded his arms for emphasis.

Bob was looking thoughtful. "Didn't I read somewhere that pollination by honey bees is responsible for something like a third of all food production? If so, wouldn't a dramatic collapse in bee population result in famine?"

Lovecott chuckled. "Or so the apiarists would have us believe. This figure has to be grossly inflated. Grasses generally pollinate by wind so that covers three of the world's major food crops – wheat, rice, corn. There are very few plants – certainly no major food crops of which I am aware – that can only be pollinated by honey bees.

"Certainly there is a massive industry in trucking around bee hives, but I wonder if that's the best option. The European honey bee is an aggressive invader, which readily goes feral. It's implicated in extinctions of native nectar-feeding species around the world including some in Australia." Lovecott addressed this last point to Martin.

"And because they have been so prolifically introduced into so many habitats so fast, they have become prime targets for diseases and mites. This has ironically cut back the feral populations, so the commercial hives have become increasingly important for pollination – but we have to remember, there are many native species that could also pollinate, which have been decimated by honey bees. So it's not that simple.

"One or two seasons' crops may be substantially reduced – mostly fruits and nuts – if the honey bee disappeared but the wild pollinators would return, and the end result would be a drop in yields, but a less fragile system.

"But I take your point that commercial agriculture has made the human ecology more fragile – not less. Massive monocultures, with decreasing biodiversity – magnets for viruses and other predators. It is not inconceivable that an entire food crop could go very suddenly."

Lovecott paused, evidently collecting his thoughts – monologue obviously one of his more finely honed skills.

"Of course this isn't a global warming effect but it does illustrate how fragile humanity's place in the ecosystem is. One worldwide epidemic that science can't control that takes out a significant fraction of the world's food production, and we *could* have widespread famine."

Martin nodded. "Finish your lunch. Repeat that to the camera. Then we can relax and take in the scenery."

Lovecott smiled. "Always take in the scenery."

*   *   *   *   *

The motel wasn't too shabby – a fair step upmarket of a hostel.

Martin sat on the bed, suddenly feeling very alone. Lovecott had been yet another intense experience. How do people live in the fast lane? Isn't this one step away from manic depression – the extreme ups, that have to dump you in extreme downs?

He fired up his computer, and noted that the free WiFi was working, then checked himself. First: back up the interview. Take no chances.

Then, on an impulse, he did a search on manic depression – officially called bipolar disorder. Eventually, he decided he was OK. If he really had a mental health problem, it wouldn't be driven by external events – not entirely anyway.

Feeling better, he consulted his list of contacts. All really more of the same. Still, duty called: now the movie was sold, he had to finish it. So he made some phone calls and set up some meetings for Monday – some new, others confirmed.

He went down to the lobby and asked about car rentals. They gave him some pamphlets and a local map. He went back to his room, found a rental company with a local depot, and organized a car for later. Might as well walk around Palo Alto a bit, check out the layout. . .

He walked out to the road, El Camino Real – stretched out across the entire map, bisecting it northwest to southeast – and worked out which way it was to the car rental place. It was a longish walk, but he had a better feel for where the university was in relation to Palo Alto when he got there. Still a couple of hours to the rental pickup time. . . he walked on into Palo Alto. A street was called "Homer". Simpsons fans? He walked up it and found Emerson. OK, maybe the other Homer. Literary figures. He found a nice shop selling whole food stuff and some likely looking restaurants. Pity the others had to go back but. . . classes, homework. . .

After walking around for a bit, he headed back towards the car place.

So near, yet so far. Palo Alto could almost be on a different planet compared with Berkeley. Quiet, laid back – not the energetic university town vibe – despite the towering presence of Stanford.

Back in his room, the car parked, he fired up the computer again, and used Skype to call Angie.

Her grin was sharper than usual, thanks to the shorter-range Internet connection.

"I miss you. How many years ago did you dump me in this place?"

She laughed. "Do some homework. I'm battling a bit too but doing homework really helps."

He told her about the car. "Good," she said. "Look around the area a bit before you head back here tonight."

"Yes. I don't really need it to get to Stanford. It's pretty close. Wish I had a bike here – everyone seems to get around the campus on bikes."

After about an hour of talking, he said, "Uh, oh."

"What?"

"Forgot you're doing homework."

"Nearly done – and anyway, we always talk at least this long."

He explained some restaurant options.

"Yes, I've heard there's some nice places in Palo Alto. We don't go there that often – San Francisco is so convenient for a night out. Get something nice."

"Yeah, but no swim today and I don't want to get fat."

"Look at Lovecott. He must be about seventy and not an ounce of fat. Walks up to the hills every day, I'll bet – anyway you did that exercise at least. And I'll get Dale to haul out his camera again if you show any sign of fat. That should frighten you thin."

He grinned. "Sneaky stunt, that telling me he was gay when I was in a swimsuit pose. I hope he didn't hear you – he'll think I'm prejudiced, or something – that look of shock and horror..."

"Naah, he didn't hear. He just thought you were clowning."

*   *   *   *   *

Monday morning. He checked out of the motel and walked to the campus early, to make sure he could find all his appointments.

After he found the first two, he stopped to check the campus map, and looked up from it at a bicycle store, opening up for the day. Just as he looked that way, a skinny-looking bike, speed written in every detail, was being put on display. He walked closer.

"Checking out bikes? This one's a great buy." The store guy proffered the price tag. "Five hundred dollars even. Didn't take it down a cent to make it sound cheaper – it's such a great buy. That's a thousand dollars plus, new."

Martin admired it.

"Pick it up."

He reached out a hand. The bike seller grabbed his little finger. He took the hint, and hefted the bike, almost throwing it in the air, it was so light.

"Top of the line – carbon fiber, weighs less than fifteen pounds. You taking up racing?"

"No, I don't think so – more like a commuting alternative."

"OK, so I have something for that: Fuji Touring – twice the weight but handles pavement cracks better, and comes with a rear rack. Easy to add on panniers or packs. Great city bike – gearing maybe a touch high for real touring, mountains and so on. Over seven hundred new." He turned over the price tag. $200.

"Take it for a ride. I'll look after your backpack." He handed Martin a helmet, and unlocked the bike from its rack.

After a quick ride, Martin was almost sold.

The bike store guy moved to close. "They buy a bike, graduate, dump it on me, and I sell it to lucky guys like you."

"Oh. They can't take it with them? I'm only here today, going back to Berkeley, then home to Australia."

"Sure they could. They could pack it up in a box like the original packaging but most don't think of that – they see a big bike and don't see how they could carry it onto a plane... wait a minute. Aren't you the climate change guy on TV? Took apart that Harvard prof?"

"Uh, yes..."

"Tell you what. I'll throw in a lock and helmet, and deliver to Berkeley. Packaged to travel."

"I want to ride it in Berkeley."

"No problem. It's a lot easier to dismantle a bike than most people realize. I'll show you how. You may want to go to a dealer to set the brakes and gears when you put it together."

Martin smiled. "Deal." By the time they'd worked out delivery arrangements to the Green house, allowing that he guessed swimming would happen at Berkeley at noon, he had to rush to his first appointment.

There followed a day of talking to research groups, being taken to lunch, filming interviews and being accosted between meetings by people who recognized him.

In something of a daze, he found his way back to the motel to pick up the car. Before starting out, he sat in the car, backing up his filming onto DVDs, and called the house with Angie's phone.

She answered on the first ring. "Good day?"

"Great. But getting tired of half the world knowing who I am."

"Doesn't it have some advantages?"

He smiled. "Some. But tiring." He didn't mention the bike.

"Still planning on driving around a bit?"

"I suppose I should – see what the area's like. How's the traffic likely to be?"

"I guess not too hectic if the interstates stay clear. You should be back here in an hour if you don't get lost."

"Compromise – it'll soon be getting dark anyway so I'll take a quick look around, then head home – back in maybe an hour and a half."

After driving around a bit through Palo Alto streets, he decided he wasn't seeing a whole lot that was interesting enough to keep him from Angie and headed back. Traffic was flowing freely on the I-880 but things slowed as he got closer to Berkeley.

Eventually he reached the house, pleased to be back among friends, no: family.

*   *   *   *   *

For a change, everyone was up in time for breakfast.

Melissa smiled at the scene of everyone at the table, things looking less hectic. "Martin, what's your plan for to-day?"

"I have to take the car back. There's a depot not too far from here. After that, we thought I'd find my own way to the campus, meet Angie for a swim, get lunch with her gang. Kind of like normal family life for a change."

"One of us could take you," Bob said. "The depot presumably is open early."

"Uh, no. I want to find out how else you can get around here... everyone in a separate car... "

"Ah, so you are questioning our green credentials."

Martin hadn't thought of it that way. "Uh... forfeit?"

Bob laughed. "Oh no, you're quite right. Car pooling just doesn't work when everyone has slightly different schedules, and public transport is not quite good enough – you have an important meeting, and work you're finishing at home... you can't afford to be late for the meeting, and you don't want to waste the extra time of being early in case the right bus doesn't show – but we *should* try harder."

Martin said, "I didn't mean it that way but I suppose we should all think about these things."

Angie was watching him closely. "I think a naughty look is being suppressed."

"Who me?" Martin deadpanned.

When the others had left, Martin dropped off the car and hurried back to the house.

Soon after 11:30, the bike arrived – just as Martin was worrying about getting to the pool on time.

"We don't usually deliver. Had to get someone to mind the store. Could have got UPS to do it but... personal service."

Martin watched him take off the wheels and put them back on, and helped him pack the packaging away in the spare room – previously his bedroom, now occupied mainly by his suitcases. As the bike guy left, he locked the house, put on the helmet and rode out into the street, his swimming things in a now almost empty backpack.

Riding the bike melted away years. He was the raccoon kid again – but at home with himself. He passed familiar landmarks, and stopped a few times to check his map. It felt good to be on the road again with his own wheels.

At the pool, he locked up the bike and helmet, with the backpack also in a loop of the lock. As he emerged from the locker room he was just in time to meet Angie on her way in. "See you in the water," he called, as she disappeared.

After the usual swimming lesson, they found the gang again – back down to half a dozen or so. Things were settling back to some sort of normalcy. He left it to Angie to hold court and bounce the occasional question about Stanford at him. He felt content to lie on a towel, absorbing the sunshine, until a cloud drifted over. He sat up.

"Done basking?" Dale asked.

He grinned. "Literally." He pointed at the cloud.

"Uh oh, rainy season's almost on us," Angie grinned.

Martin stood up abruptly.

"What's the matter? You can hide in a building until I go

home." Angie lay back lazily on her towel.

"Angie, I'm exploring non-car travel, remember."

"Oh?" one of them asked. She explained his little project to the gang.

"Gotta go, love you." He rushed off.

"That's one mystery solved," said Peter.

"What?"

"Why he hangs out with you."

He expertly ducked a solid cuffing.

* * * * *

Back at the house, the rain was coming down steadily. Martin had the bike in the bedroom for company. Lucky we got home in time. He patted the frame.

Better do some work, I suppose. He started up the computer and reviewed the Stanford clips, and made some notes for editing. Then he looked at his ticket back: less than a week and he'd be home. Whatever that meant. Home was ...where he was with Angie. PhDs – best to get that sorted now.

He stretched out on the bed, contemplating, the gentle sound of the rain on the roof reminding him of rainy days in his father's house. The unhappy kid who talked to birds, the raccoon. The rain had been one of his few friends. Such a soothing sound.

He woke up with the light fading, to the sound of the front door opening. He heard rapid footfalls up the stairs. Angie burst into the room. "Aha! Caught you in the act! In bed with –" she stopped to turn on the lights – "a bicycle!"

She dived into bed and grappled with him, in what rapidly turned into an impromptu wrestling match. "You shameless hussy!"

"Uh, isn't a hussy female?"

"Yeah, but sexist language – if you can be a babe, you can be a hussy."

"And you can be a wild animal tamer – traditional male role."

"So you're tame now?"

Sotto voce: "It can hear all this." He gestured at the bike, and she took advantage of the move to get a better hold.

Aloud: "Why, is it under age?"

"No, it mustn't know I can be tamed so easily."

"Easily?"

Another round of wrestling ensued – "Walk out on me for an inanimate object, will you?" – interrupted only when a throat was cleared at the door.

Bob was looking on in amusement. "So you now have a rival and don't like it?"

They all laughed. Bob went on: "You know, we should all get bikes."

*   *   *   *   *

Stanford was a big anti-climax after all the excitement, especially as it marked the end of the business part of the trip.

The next few days had their memorable moments – the new family routine of biking, getting caught in the rain, working through PhD options that pointed increasingly to Sydney, but the time flashed by.

It seemed like seconds since the walk in the Stanford hills, when they were all back at San Francisco Airport, yet years since he'd stepped off the plane in San Francisco.

Martin endured tearful hugs, thinking ahead to the lonely days back in Brisbane – shopping for a nice computer, the work on the movie. Should he invite Julie to help again even though he had his own gear now? What will it be like with dad, now that things weren't so distant anymore?

An hour and a half to LA. Fourteen hours to Brisbane. Time, distance . . . how do these capture the whole concept of travel? Only a few months ago, he was a different person, a being from a different universe.

17   No Rewind

BRISBANE WAS WARMING UP, becoming muggy – not quite the height of summer. That would develop through to February. Martin was desperately looking forward to Angie showing up. But there were loose ends to take care of. On a Sunday, dad should be up and about by nine – too early to have gone anywhere yet – should catch him at home.

His dad picked up the phone on the third ring.

"Hi, dad."

"Marty! Good that you're back. We must get together."

"Yes, dad. I was just thinking. You never met Julie, me being such a brat and all that –"

"Don't judge yourself. I made things pretty hard."

"Never mind that. I got to thinking, after she came up in conversation, the raccoon thing and so on, that you might actually quite like her. And I have to see her family some time. It will be hard, but it's the only way we can be friends."

"So, not an idiot like your old man – your mother and I really could have been friends..."

"Who's judging themselves now? Let's bury all that, OK?

"Look: my place is not really big enough for a family party. Could we invite them over to your place? Maybe next week end if everyone's free?"

"Of course. You know me: no social calendar. Tell me when they can make it. And how about later today? Are you free? I haven't been to a movie in ages."

"I don't know what's good right now."

"Me neither. Let's just go."

Martin smiled. "Yes, dad. It's about time a boy let his dad take him to the movies. A bit of relaxation before I go out and buy my editing gear."

"You know, son, that place of yours is not that spacious. Why don't you set up your stuff here? You could use your old room."

Sensing that his dad was not purely being practical, he said, "Good idea, dad, It'll give us more time to reconnect. Do you have time to go shopping with me on Monday? I could do with some help on the heavy lifting."

* * * * *

"That wasn't much of a movie, was it dad?"

"Nope. But good to get together anyway.

"How about we go home for a beer?"

"OK."

His dad drove contemplatively.

At the house, the older man found a couple of glasses, and poured the beers expertly.

"To family," said Martin, and they clinked glasses.

They sat for a while, sipping their beers.

"You know, Martin, I've been thinking a lot about what you said about your mother, you know, orientation not a choice and all that.

"I wish someone had told me all that. Or, maybe, someone did and I didn't listen. I don't know. How did you find all this stuff out?"

"Julie. She had all kinds of friends. Everything except conservative engineers." He grinned. "I was the odd one out – the token geek.

"This one pair who used to show up at parties kind of shocked me the first time. I'd never seen two guys kiss. But after seeing them a few times, it just seemed kind of natural, you know – Julie and I were a matching pair, or so I thought at the time, and so were they: Damien and Jeremy.

"I didn't say anything about this – I was too shy to talk most of the time anyway, but you pick up things. And you can't help being curious."

"Did you...?"

"No dad. I could hardly climb the ladder the right way, I had no clue how to climb it upside down."

"Eh?" His dad was really puzzled now.

Martin painted the scene ...

* * * * *

It was the last day of exams. Results would be out soon, but everyone pretty much knew where they stood. By the time you get to your final semester, you know the score.

So it was party time.

They went to a house – one of many Julie had dragged him to for the last year and a half.

Soon, the party was going apace. Booze was flowing freely, music was loud, and various chemicals were doing the rounds. He couldn't take the smokey air, and went outside. The house had a wraparound verandah, and he stood at a corner, taking in fresh air.

After a while, he realized there was someone else around the corner, not two paces from him. It was Damien.

"Uh, Hi." Damien looked his way – it seemed he had also not noticed there was someone else there.

"Oh yeah. One of the straight boys."

"I dunno, I mean do we have to label people like this? Most of us don't spend our whole day shagging."

"Good question, straight boy. If you walk down the street with your girl, does anyone even notice?" Damien was not entirely sober. Martin had never heard him talk like this before.

"I'm sorry. I'm just trying to understand."

"What's to understand? You don't choose to be gay or straight – or did *you* choose? Can you remember the day?"

"Uh, no."

"Yeah, well exactly."

"Why does this whole sex thing have to be so hard? I mean, before I met Julie," he thought of an analogy from his gaming days, "it was like trying to climb a ladder to reach something nice, and every time you thought you were there, it moved. And you had to start climbing again."

"Let me tell you what, straight boy. You had it easy. You were climbing on the top of the ladder. What it was like for me was more like climbing underneath. And every time I had the hang of it, someone would grease the rungs, or stomp on my fingers."

Jeremy's voice floated out from the smokey house. "Damien, lovey, what are you up to?"

Damien headed back inside. "Straight boy trying to be nice," he explained, his voice disappearing into the party sounds.

Martin wasn't sure how it was meant. He was staring into the growing dark, when Julie eventually found him. She was a bit wobbly. "What are you up to, lover boy?"

"Nothing. Can't take the smoke."

"Yes you can't." She evidently remembered the various attempts at getting him to try dope and his consistent inability to comprehend that it was *anti*-nauseal.

"You should celebrate somehow. Remember, I start work on Monday." She headed back inside.

He thought of saying, "But I don't," but didn't. She was gone before he could form the words, anyway.

The party wound down. He went to look for her. She was nowhere to be found.

He didn't know it then, but that was the beginning of the end.

*   *   *   *   *

His dad looked thoughtful at the end of the story.

"So that's when she walked out on you?"

"Looking back on it, yes. She saw the advertising job as selling out, and she'd dragged me away from my big dream of gaming."

"My poor boy. We were all against you."

"Dad, that's exactly the way I saw it but, you know, you all did what you thought was best. I mean, no one helped you with understanding mum's problems – no one you could relate to."

"That's true. I used to go to church every week, until it happened. Your mum wasn't so keen. She would go sometimes. But afterwards, some of the old fogies said such terrible things about her. I walked out and never went back."

There was a long silence.

"Family? Why didn't you ever talk about it to family?"

"Same thing. We were from farming stock – very conservative. I was the first to move to the city. They also slagged her off. I guess they thought that put them on my side. I can still see the looks on their faces when I told them they weren't family anymore, unless they took that back."

"Dad, I am so pleased that you stood up for her."

"Yeah. But I didn't tell her."

"No rewind button – it's past. You told me, and that's the best you can do. And *her* family?"

"Also took it hard, but accepted it, in the end. Not *quite* so conservative. But they took her side. Didn't want to know me."

"You know what?"

"What, Marty?"

"I really do think I need to meet this person. What was her name, my mothers other, ah, partner?"

"Geraldine. Geraldine Simmons. Are you sure. . . ?"

Martin was sure.

* * * * *

Looking at Julie didn't hurt as much as before – but the old
chemistry wasn't quite dead. Martin let them in, and his
dad grabbed James and steered him to the back deck, talking BBQ. James towed Donnie after him. No signals were
necessary: the grown-ups understood this was the time for
Julie and Martin to reconnect.

He led her downstairs, and to the back yard, out of sight
of the house, behind some tall shrubs.

Julie broke the ice. "How's... Angela?"

"Great. Just champing at the bit over the bureaucracy to
get a study visa. After all the unis were falling over each other
to accept her, they had to get out a formal letter to her, which
took forever, then it turns out she needs a chest x-ray... one
thing after another. On the web site, it says five to ten working
days, but it's kind of hard to work out *which* five to ten days."

"It's wonderful to see you looking in such good shape."

He pounded his pecs, gorilla-style. "Swimming."

She laughed. "That too. I mean feeling good about yourself."

He grinned. "It's been a long, hard climb. But I don't
look back – not a lot anyway. And you?"

"I'm doing... OK. Donnie is a handful, James is really
sweet. Did I tell you? He's an architect. Isn't your dad an
engineer or something?"

"Civil. Big stuff. Tunnels, airports, dams. And no, you
didn't tell me anything about James. We were pretty busy
editing a video last time we met."

"Is that why you kept me away from your dad – not green
enough?"

"No, no. I was a troubled kid." He told her the story about

his mother. She gave him a gentle hug.

"I'm so, so sorry. I had no idea. I thought... I don't know what I thought. I had no right interfering with your life like that. I could have done so much harm."

"I liked it a lot at the time. You were the best thing in my life – but we aren't supposed to be looking back. I hope you'll like Angie."

"I hear you guys are calling each other 'partner'."

"It just sort of happened. Everything we do together is so good, even the fights are world class." He grinned.

"Are you going to change it to 'married' some time?"

"I don't know. I kind of feel it will also sort of happen when it's good and ready."

"I hope so. Listen: I may have made a stupid mistake with you, no, lots of stupid mistakes, but, you know, I'm always here if things get screwed up and you need someone to talk to."

"Thanks." He hugged her back, and they risked a kiss. She used the opportunity to explore his swimming build, and reluctantly let go.

She shuddered. "That was too good. Let's go back inside before we get each other into trouble."

He smiled, and they held hands for a couple of paces.

Back inside, Donnie was getting bored with grown-up talk. James and Truscott senior had obviously hit it off well, and were arguing urban development ideas, building styles, freeways versus tunnels, cars versus public transport...

Martin interrupted. "Dad, I think you should get to know Julie."

He got down on his haunches. "And who's this?" He asked, taking Donnie by the hand. "I'm Martin."

Donnie's eyes grew to the size of saucers. "Wow. Mom says you're superman. Can you fly?"

Everyone else but Martin laughed. Martin looked very serious. "No. Can you?"

"Don't be silly. I'm not superman."

Martin laughed. "I'm not used to playing with little guys like you. Can we play outside? I have a special place I used to play in when I was little."

Julie and James nodded consent.

He led the toddler by the hand to the bottom of the garden. Donnie said, "Mom says you can slay dragons and fight off mon-sters."

"Donnie, do you know what 'exaggeration' means?"

"Extra-geration? Isn't that a grown up word for fibbing?"

"Nearly right. It really means making things seem bigger than they really are."

"That sounds like fibbing to me."

"Donnie, I am not going to argue with a superior intel-lect."

The little boy smiled at the grown-up words.

"Not even a little dragon?"

"Donnie, I used to slay dragons when I played computer games."

Donnie's eyes lit up. "Have you got games? Mom won't let me but my friends. . . "

"Sorry, Donnie. I don't like playing computer games any-more."

"Oh. Is it true that you are the boss of the Eff Bee Eye?"

Martin laughed. "What has everyone been telling about me?

"Let's stop talking about that. Let's sit under this tree and keep very quiet. I want to show you something."

Martin sat on the grass, one knee raised. Donnie sat against his raised leg, looking around expectantly.

Martin put a finger on his lips and pointed.

A little black bird with a long tail landed near them, and started hopping around. Soon, another arrived. They hopped closer, looking curiously at the two people.

Martin whistled gently. Donnie said excitedly, in a whisper, not losing the moment, "You can talk *bird*."

Martin shook his head, and whispered back. "No. This is just my way of saying hello." He whistled again, and the birds hopped nearer, stopping to inspect them carefully. One of them got really close, and Donnie reached out his hand. Martin held it back. "Look, don't touch. We are so big and strong. If they think we are going to touch, they'll fly away."

This scene continued for some minutes, the birds now so close that they could easily be reached. Then a noise from beyond – and they were gone. The pair of them looked up. "Mummy! You frightened them off."

"It's OK Donnie." Martin reassured him. "They wanted to go away. They'll play again another time. Would you like that?"

Julie leant down to Donnie, and said, "Your daddy has your lunch ready. Run along, and I'll bring your new friend with me."

Donnie scooted off, and over his shoulder shouted, "He's not superman! You're a fibber!"

Julie watched him run off.

"How long were you watching us?"

She turned back to Martin. "I don't know, several minutes. The two of you were having such a good time." She paused, and looked at him, an unfathomable expression in her eyes. She said quietly, "Don't you wish he was yours? I saw the two of you together, and you looked so much like a dad."

He gave her a searching look, and said, eventually, "I am so glad he's yours."

She started to speak, and he interrupted. "Don't even think of it. The boy you used to know was no way equipped to be a dad. Let's just be friends now. Like I told my dad, life has no rewind."

Back at the house, food was ready, on the back deck table. For Martin and his dad, there was salmon; a variety of traditional BBQ items for the rest, without food issues – sausage, steak, onions, salads, rolls. Everyone besides Donnie, who was already busy, helped themselves.

Conversation continued over food. Afterwards, ice cream appeared in several designer flavours. Said Truscott senior, "Last time my son was here and I needed ice cream, I was remiss."

"Not this time, dad." Martin picked up a low-fat gelato. "Even allowed for people with a cholesterol problem."

Julie gave Donnie a small serving of several flavours, and they all awaited his expert opinion. He looked up, several colours on his face. "All good." Thus reassured, everyone else took a few flavours.

Julie noticed that Martin kept checking his watch. "What's

the matter, do you need to be somewhere?"

"No, but it's getting close to my time for talking to Angie. Would anyone object if I retire to my room to talk to her?"

Julie said, "Your room?"

"Yup. Staying with my dad for a while, and doing my editing here."

"Any chance I can meet her?"

"Me too," added Donnie to laughter. "Is *she* superman?" This was even more successful.

Martin said, keeping a straight face, "No, but she's stronger than me."

"Not right."

Martin contemplated, then nodded. "I think it should actually be 'Stronger than I'."

Donnie ignored the grammar lesson. "I don't believe it."

"It's true. I can't do anything unless she lets me."

"More grown-up fibs."

Martin focused attention back on the main issue: introducing Angie. "Dad, you too – it's time you got introduced to Skype."

"OK. Or use the phone."

"Thanks, dad. I don't really like the Skype echo and all that, but I like the video."

They all followed him to his room. Julie looked at his equipment. "Don't need me anymore."

"Don't be silly. I'd love to have you give a professional opinion and help with some of the details, if you can spare the time."

She could. James nodded.

Funny after all this time, this was the first time she'd seen his room in the family house. And it once again had a big, top-of-the line computer in it with an industrial-strength graphics card – one that would be good for games, yet set up for the kind of work *she* did every day. But for the kind of project she would have related to back then.

He pushed the ironies aside, moved the keyboard of the big machine out of the way, pulled out his trusty MacBook, and set up the call. Angie was waiting, as expected. Several faces clustered together around the camera – everyone except Donnie, who was way too short.

"Hi, Ange. Some introductions. I don't think you've seen my dad before. And this is Julie, and James."

He lifted up Donnie. "And their very own Donnie."

Greetings were exchanged, then Julie said, "Come on guys, this place is too crowded."

When they were alone, Angie said, "I didn't expect that."

"I did say we would be doing this party."

"True. But I'm so used to seeing you alone. How did you take meeting her family?"

"OK, actually. I was scared that I would crack up when I saw Donnie, you know, feel he should have been mine. But we really got on well. I showed him my little bird corner."

"And Julie? How did you handle her?"

"OK. I think we can be friends. James was really good about it – he let us have some time together."

There was a silence.

"I have a confession."

"Yes?"

"You remember that time you made me cry?"

"How could I forget? You may not have noticed because you ran away from me, but I was pretty low too."

"Yeah, well I lied when I said that was the first time anyone ever made me cry."

"Oh."

"It was the second time, and the first time was you too."

"When did I do that? I thought everything was unbelievably perfect up to then. A little too perfect. I kind of feel better now we've had some battles."

"When I reminded you to ask her to help with editing. I hardly knew you except for that magic weekend. I didn't know she'd developed a family. It was kind of easy in Boston, when we were together to offer the idea. But when I reminded you – I mean, I could have tried to put you off the idea, or had some other option, but I didn't. When we ended that call, I though, 'Oh crap. What have I done? What if they get back together?'"

"Why did you do it then?"

"You wanted that thing done so much, I just couldn't stop myself. Tell me straight: what *would* you have done if she'd wanted you back?"

"Become a mormon?"

"You *dog*! You reduce me to tears twice and – hey! Where did you learn that naughty grin?"

"Mainly from you – and last time I checked, my very own, loving personal biologist certified me as at least a great ape."

"Jeez. I'd better be careful what I teach you in future." She grimaced in mock horror.

"But ... " he looked serious.

"What?"

"My darling, you never had any competition since you touched my shoulder in the Blue Parrot."

"Do you really mean that?"

"Have I ever lied to you?"

"No. No, and I should learn that from you." She smiled – something more introspective than her impish grin.

"How's your visa going?"

"Nearly there. I think I've cleared the last obstacle."

"Great. Otherwise we would have had to get married."

"Oh. Should we?"

"Just to get you in the country?"

"No. Because we want to. Do you?"

*   *   *   *   *

Shadows were lengthening by the time Martin surfaced.

"Sorry guys. We could have kept on talking for hours."

"You did," his dad pointed out.

"Hours more."

Donnie was asleep on a couch, and his parents were looking an hour or so past ready to go.

"You guys should just get married," Julie said.

"How did you guess?" Martin missed the exact sense of what she said.

His dad said, "Is this something someone forgot to tell me yet again?"

"No, dad. We just decided, and this time you're the first to hear, unless she woke up her parents really quickly. It's about midnight over there, and she said they went to bed already."

There was a moment of silence, then excitement, as everyone tried to give him bear hugs and share high fives, with some ensuing tangles arising out of incompatibilities between the two manoeuvres. Martin explained how it started from the very practical thing of the study visa, and ended up as 'what are we waiting for?'

George Truscott smiled. "While we're talking practicality, there's a pretty good migration agent I've used a few times when I had to import talent. If anyone can handle the details with no glitches, it will be him."

Martin smiled his thanks. No words passed for a minute. "Dad, I want to do something very untraditional."

"What's that, Martin?"

"I want a best woman, if Julie would like to do that."

She nodded slowly, unable to speak, and put a hand on James's arm for support. He nodded too.

"That's as untraditional as you want?"

Martin grinned at his dad's question. The old man had moved a long way. "We'll have to work on the details with Angie. But that's the only thing I specially want. And no church. None of us do church."

Donnie woke up. "What's happening?"

Julie picked him up and pointed him at Martin. "Superman's getting married."

"He's not superman," he pouted – "but are we all invited?" Everyone laughed.

*   *   *   *   *

He rang the doorbell. A silhouette appeared in the frosted glass pane of the door. The door opened slowly. A short, stockily built woman with greying hair, large earrings and an unfathomable expression confronted him.

"Why did you want to see me?"

"May I come in?"

"I suppose so."

She stepped aside.

He offered his hand. "Martin Truscott."

"I see we are being formal. Geraldine Simmons." Her handshake was firm, but without warmth.

"No, polite.

"You were the last person who really knew my mother. My dad hid her from me. He meant well, but he was wrong."

She softened a bit. "Sit down. Would you like some tea?" He nodded.

When she returned from the kitchen with a tray with cups, milk, sugar, a teapot and biscuits, he was staring at a photograph, with eyes that stared back at his, as if a reflection – so similar were they.

"Yes, that's her. A treasured possession."

He nodded.

"Just tell me about her. I hardly knew her. The visits were so short – then I lost her . . . "

"She was a great wit. Being with her was always fun. Always a ready smile. It was so hard, though, when the subject of you and your dad came up. Wait here – I have something for you."

She went away for a few minutes, and returned with an old-fashioned photo album. As he took it, she made tea to his specification – no sugar or milk – and handed it over with a couple of biscuits on the saucer.

He opened it, uncomprehending. "Who's the baby?" Then realized it was a stupid question. There was only one baby. He turned the pages. As the baby grew, it looked a bit more like him. The pictures stopped at about the toddling stage.

"Your tea's getting cold."

He drank it fast.

"Keep it." She gestured at the album. "I should have given it back, but –" she hesitated – "I was selfish. It's really yours."

He was looking at the picture of his mother again. "Could I borrow that? I can make a copy."

She smiled for the first time. "Of course. Just don't be too long."

"I was also wondering..." he felt awkward.

"Yes? Oh, I suppose you are wondering how we were infected. Very silly thing. We were holidaying in Amsterdam – one of the first places you could be openly gay and just have fun. Body piercing was a big thing there. We went with the flow – didn't think there'd be a problem in a first world country. But there was.

"I got sick first – bad coughs, we thought at first it was just out of season flu. Then she started to get all kinds of strange ailments. So we both got tested. It was a terrible shock. It had to be the Amsterdam thing – we eliminated every other possibility.

"We both went on the drugs. Terrible stuff in those days. Her hair fell out, I was constantly nauseous. But they adjusted the dose, and I eventually started to come right. The drugs are much better today. But she..."

Martin went over to her and gave her a hug.

She held him tight. Eventually, she pushed him away, and said, "I never expected that."

"Well, I think it was very unfair that I lost my mother, but people get divorced all the time. The way it happened between us all was a flaw in society."

She nodded. "You know, when I saw you on TV, I was in tears. Your mother would have been cheering for you. She was such a fun person, yet she had a stern streak. I see both of those things in you."

He smiled tightly, keeping emotions in check. "I have something else. I'm getting married. It would mean a lot to me if you could be there, representing her."

"Your dad?"

"There's lots of time. We can work on him. I've already started."

"You don't know how bitter the rest of his family was."

"I don't know any of them. He walked out on them, out of his church too. He couldn't take them saying bad things about mum. And I won't talk to them either. He didn't deserve that, and she didn't deserve it either. Sometimes, you are just wrong, and reconciliation can only happen from your side."

"I really didn't know. The poor man. And we thought we had it hard. My family was OK, and her family eventually accepted us, even if they were never that comfortable with the idea. I even still see some of them sometimes."

"Do you think you could introduce me?"

"Of course. But the hard work will be up to you. And them. I can't invite them to your wedding but...I would love to be there. But only if you are really sure it won't hurt George. If it would, send me a video. You're pretty good at making those."

As she was showing him to the door, he said, "One more

thing. We aren't being totally conventional. No church, because of what they did, and anyway, my partner and I never did church, so it would be hypocritical.

"And we're not sticking to gender conventions –" he paused for dramatic effect – "a woman in the 'best man' role."

She laughed. "Exactly like your mum. Such a wicked sense of humour. You nearly had me there. Who are the two lucky girls? I'm sorry, all the emotion and everything, I should have asked."

"Angie is my partner – met her in Boston, spotted me reviewing my infamous interview. She lives in California: should be here really soon, once we get her visa sorted. Julie was my girl at uni, and I never had any real male friends. 'Best woman' seems to fit her pretty well, even though she caused me lots of grief."

She opened the door. "What was that song, 'fine line between pleasure and pain'?"

He gave her another hug before going out. "Did you forget about the HIV?" she asked, almost as an afterthought.

"No. I know you can't get it from contact.

"Look after yourself."

18 Family

SKYPE TIME AGAIN. Martin stretched, backed up his editing, and switched computers. Angie was waiting as usual. He went straight to business.

"I had a chat with the migration agent. Bottom line: your study visa is nearly ready: get it. There's a lot of complications with spouse visas, I guess to prevent abuse. He can handle it all for us but the bad news is ... two year wait before you can get permanent residence. And that means you pay full tuition for your PhD for two years."

"Oh. So we don't have to get married, then –" she went on quickly – "look at that face. Remember, we both decided 'want to'. It's so much nicer if there is no hint of 'have to', isn't it?"

He grinned. "You really had me there."

"How's the editing going?"

"Good. Julie is coming over tomorrow to help work on some details."

"Great – I hope we can talk this time. I'd like to get to know her. You met my whole gang, and I don't know any of your Aussie friends."

"No rivalry?"

"She's had every opportunity to rip you away from me, and look at you – almost in tears when you thought I said we needn't get married; besides, she's on the team."

Martin's routine had settled to a trip to the pool, each time cursing the public transport inconvenience from the house compared with his place, a few hours of editing, going for a walk outside, a few more hours of editing, visiting his bird friends, a little more editing, then Skype to Angie. Then still more editing until he was too tired to focus ...

It would have been much the same in his apartment, except the brief conversations with his dad over breakfast and dinner – a growing pleasure now that the differences had been bridged – and of course, the birds. That night, he discussed progress with the visa with his dad. "I can't set a date yet, but she's going to be here pretty soon. We're shooting for the first flight she can get after exams. That's in two weeks."

"I'm looking forward to this almost as much as you, Martin. Can you think of something nice I could get her, to welcome her to the family?"

Martin thought for a bit. "Yes, I think there is."

He explained the iPhone thing. His dad nodded. "I'll look into it. Would the Apple dealer in the city where we picked up your machine have them?"

The next evening when Martin got back from his daily commune with the birds, there was a package on his bed. He opened it. Not one, but two iPhones.

"Dad! I didn't expect this."

"And who exactly did you think she would spend all her time talking to?"

*   *   *   *   *

Martin unlocked the door. The feeling that the old house was home again was good in a way, but this really was his own place – covered in dust though it was. If Angie was to be here soon, it had better be in good shape. He contemplated the bike, mostly unused since he got back – best leave that somewhere conspicuous, an old friend to greet her...

He found the vacuum cleaner, and set to making things right. When the dishes were sparkling in the drying rack and he couldn't find anything else to clean, he went shopping. As he got back with some beers and a few groceries, he found three people about his own age camped on his doorstep.

They had that sincere, idealistic look he remembered from some of Julie's friends. Sure enough, one of them introduced them: 'Hi, Martin Truscott?' Martin nodded. "We're a delegation from your local branch of the Queensland Greens. Can we talk?"

He smiled. "Good to see you. Here – hold this." He handed a shopping bag to the leader of the troop, so he could find his keys. He let them in.

Once inside, the leader offered names. "I'm branch convener, Hugh Sidebottom. This is Janice Sidebottom, treasurer, and Dave Miles." Janice had some papers with her – forms and pamphlets.

"Sit down – guys. I've shopping to put away. Including some beers. Still cold. Would you like one?" They liked.

He brought back four opened bottles.

"So, to what do I owe the pleasure?"

"Well," said Hugh, "we really admire what you're doing

about climate change."

He held up a hand. "Now, wait. That thing on TV happened because Schoor is an asshole. He set it up, I knocked it down. CO_2 is still pumping out just as fast."

"Exactly," said Dave. "And we need people with credibility to speak out and stop it."

"Credibility? You'll have to judge that by my finished show."

It was Janice's turn. "We have no reason to think it won't be good. Anyway, you have a public presence. If you show up for anything, the press will be there."

"Don't I know it." The scene outside the Greens' house – the Berkeley Greens, not the Greens Party, stupid thought – swam before his eyes; the swarm of cameras. "Anyone would think I was Princess Di. Still alive," he added fast.

They all laughed.

"Anyway," Hugh went on, "the federal election has to be called soon..." Martin had a cold feeling in the pit of his stomach. This could only be going one way. "...and the Greens have never won a lower house seat, except once in a bye-election, and we lost that one, next election."

They could see he was looking doubtful. Janice took over again. "If we can make at least ten percent nationwide, and pick up one or two lower house seats, it will change the whole debate. Make the big parties realize they can't be complacent."

They allowed him some space to think.

Slowly, he shook his head. "Guys, I'm one hundred percent with you. I see you brought a bunch of forms. I'll happily sign up as a member. Put in money, campaign, walk the streets, give speeches if it helps. Here, Sydney, anywhere.

"But I've been in the goldfish bowl already, and I'm not sure if I'm ready for this, full-time. I'm making plans already to do a PhD at UNSW, and it's really important stuff – ice sheet modeling in the Antarctic. The NASA people are convinced that a major ice melt is almost certain, but they don't have the detailed modeling to back it up.

"And also, I'm getting married, and I want to have time for Angie." Selfish though that sounds, he didn't add.

The conversation continued around election issues, campaign strategy, and the like. Martin filled in a membership application while they were talking, and handed it back to Janice. Maybe they're trying to interest me, and they are... Eventually, he said, "I am really sure that I have to focus on the ice thing. It's just too important."

They stood up, disappointment showing in every detail of their body language. He noticed that they had not picked up the nomination form.

"Guys, listen, if you can make the voters feel half as bad as I do, you'll have a clear majority next election." That cheered them up slightly.

As he showed them out, he added, "Angie and I always make the big decisions together, OK? So I'll talk to her as well. She'll be here in a couple of days. Not holding out hope or anything, but I am sure she'll help with any campaigning."

As he closed the door, he sagged to the floor. Man, that was hard. Like cows' eyes to a vegetarian. And I used to think being unpopular sucked.

$$* \quad * \quad * \quad * \quad *$$

Brisbane Domestic Airport, Friday late afternoon.

Angie's flight from Sydney was due in, only a few minutes late.

Two iPhones in pockets on either side of his pants, one waiting for its rightful owner...

If her trip via LA was anything like his, she'd be just about dead. He found the arrival gate with a couple of minutes to spare. People started to stream out. He peered over the taller ones, making sure he didn't miss anyone. Surely even if he missed her, she would spot him. The last passengers walked past, all the other meeters with them, and she wasn't there. The gate closed in preparation for the departing flight. He ran in a panic to the luggage carousel. The bags for the flight were only just starting to appear. She couldn't have found hers already, and gone off somewhere. As people picked up their luggage, it was increasingly clear that she was not there. Worse, he saw at least one bag was tagged from LAX. Could there be another flight from LA that this one connected to? He couldn't see the flight number before the owner of the bag grabbed it, and headed for the exit.

He went over to a Qantas service desk.

"My partner was supposed to be on that flight from Sydney" – he indicated the number over the carousel. "She was connecting from LA. Could you find out if the LA flight was delayed?" He gave her the international flight's number.

"Let's see." She tapped at her keyboard. "No, landed on time."

"Can you check if she somehow missed this flight, maybe made it to a later one: Angela Green?"

"Sorry. Privacy and security. We can't give out that kind

of information."

Seeing his look of abject misery, she typed again. "Let's just say you would do well to wait at gate forty-two for the next flight from Sydney."

He made as if to grab her and kiss her, then thought better of it, and sprinted for the security checkpoint.

"Wow," said the person at the next counter. "Did you get his phone number?"

"Look at that speed. I'd say someone else already has his number."

At gate forty-two, the Sydney flight was due in twenty minutes. His phone chirped. He picked up the call. It was his dad. "Martin."

"Dad, she's not on the flight. I'm hoping it's the next one."

"Yes, she called from the airport. Seems she offended the immigration people somehow. They took forever questioning her, but she should be on that flight. I gave her your mobile number, but she had to run to catch the flight, out of coins."

"Thanks again, dad."

"For what?"

"The phones."

Twenty minutes had never taken so long to pass. He kept thinking of the stupid gate number forty-two. The answer to the fundamental question of life, the universe and everything. "The question? Why is this the same number as McCarthy's house?" He realized he'd said the last bit out loud – there were no *Hitchhiker's Guide* fans at the gate obviously. His neighbours looked at him as if he was crazy. He paced up and down silently. Let them think what they like.

Angie was the third person out of the plane, sprinting past everyone else, almost making it past the business class bunch who were let out first – her backpack battling to keep up.

Martin stood in her path. "Hey! I'm supposed to bowl you over. This is *my* home airport."

"Ours now."

He led her towards baggage claim. "What happened to you? Dad said..."

"Oh, it's so silly. You know on the entry form, it asks if you have any criminal convictions?"

"Yeah – do you?"

"Well, no, but I pointed that out to the immigration dude, and innocently said, 'I didn't know this was still required.'"

Martin laughed.

"I also thought it was funny, but he didn't, and they dragged me to an interrogation room and asked all kinds of questions. They only stopped when your name came up.

"It was like, 'Oh, crikey, he's a national treasure, why didn't you say so?'" Her attempt at faking an Aussie accent was straight from the Simpsons. But Martin's grin faded.

"Oh. First, people think I'm in charge of the FBI, now it seems I'm in charge of Australian immigration. Power corrupts. Hope the media doesn't get this one." They arrived at the carousel.

"Speaking of which..." he told her about the Greens – the party – as they waited for her stuff. She was leaning on him, obviously tired, but also contemplating the whole thing. She had two large suitcases, and they wheeled them to the train without further conversation.

On the train, she said, "It could be kind of fun to go into politics."

She put a hand on his knee.

He shook his head. "I don't know. I mean, I think I could do it, but the ice project..."

"You're right. That work is really important. And I'm not a citizen, so I guess I couldn't run either."

"That takes years to set up even after you get permanent residence and even if it didn't, if you spent half your life in Canberra, what kind of team would we be? I really want to stick with Wilkinson, even if there are PhD options in Canberra."

"A pity in a way, though. I think you would give their campaign a big lift. Hey! I have an even better idea..."

He nodded as she explained. "But first, we have to get you home."

In twenty minutes, they were at the Brunswick Street station, in Brisbane's Chinatown.

Martin tried to point out landmarks as they headed towards his place in Fortitude Valley, but gave up. Steering her in the right direction was hard enough on its own. It was nearly 6 pm when they got there – well past midnight in California.

"The Truscott theory of jet lag says you need to keep awake for another three hours."

"Hey! An old friend." She patted the bike. "Used it much?"

"No – I was waiting for yours to show up so we could bike around together. I kind of felt it would feel lonely out on the streets without company."

She nodded, and pulled some documents out of her back-pack. "My other stuff will be showing up soon. But first, a shower," she said, putting everything down.

"No, first, your surprise." He hauled something small from a pocket and held it behind his back. "Close your eyes. I was going to give you this at the airport, but kind of lost the moment when your flight was delayed."

There was something cool in her hands. She opened her eyes to see a shiny new iPhone. "How did you know?"

"Well, it's from my dad. He got us each one. To stay in touch. And play music and..."

"How did *he* know?"

"From me." Impish grin.

"How much math *did* you do? Reducing a problem to a simpler one." She grinned too. "Of course – the Berkeley pool conversation. Can I hug you instead of him?"

He didn't mind. "Let's see how we both fit into my shower. Can't have you falling asleep in there."

"Hey, another excuse to share showers."

"Who needs excuses in our own home?"

"Our own home. That sounds so good. Wait," she suddenly looked concerned. "I want to unpack one thing. I really hope they made it in one piece." She opened both suitcases. "OK, in here," she said, after scanning the contents, and picked one. She carefully rolled back the top layer, and pulled out a box, which she opened. He peered over her shoulder, as she took the contents out. "Our glasses."

"Ours?" They were the glasses from Boston. "You weren't thinking we were married already in Boston, did you?"

"I don't know. It wasn't conscious, I suppose. I wanted to buy something memorable to take home, and caving in to the iPhone fashion didn't seem so memorable – not after all the fun we had. And of course, these were linked to you."

"I was wondering what happened to them. I didn't see them in California, but what with all the excitement forgot to ask. I'm so ignorant of this sort of thing. I kind of thought a good glass cost a couple of bucks, ten at the outside. We talked about this at the pool, then I forgot about it. I looked up Riedel when I got home from California, and kind of picked up that they are pretty expensive. How much *did* you pay for these?"

"Uh, a bit over a hundred bucks..."

He put the glass down reverently. "Wow."

"You must admit, the wine was spectacular in them."

"You're not saying you spent a hundred bucks on the wine too?"

"Well, no. Only eighty."

He grinned. "I suppose you came out ahead. You spent less than the cost of an iPhone, and you got one anyway."

The shower lasted longer than it should have – eventually Martin said, "We're supposed to be saving water here." As he turned the water off, he added, "Anyway, you look a bit more awake. Feel up to taking a look around outside?" he asked, passing her a towel.

"Yes, I think I may actually pay attention a bit – things got a bit hazy after the train. We did decide not to go into politics didn't we? Oh yeah... my big idea..."

*   *   *   *   *

The next day, they both woke up early, about 6 am. Martin said, "My turn to do room service." He returned with two cappucinos, and a pair of freshly warmed croissants. "Anything else you'd like? Orange juice?"

"Mm," she said through a bite of croissant.

"I'll take that as a yes." He fetched it.

An hour later, they were sharing parts of the big fat weekend editions of *The Australian* and *The Sydney Morning Herald*, spread out on the bed, with fresh coffees on the side.

Angie was looking puzzled. "So explain this to me. There's the Labor Party which I guess is sort of like the British one?"

"Right. Socialist history, now more or less centre-right – not much in bed with unions any more."

"And this other coalition? Liberals? They sound pretty conservative to me. Who else is in the coalition?"

"Well, the whole thing is pretty simple really," he explained, adopting a professorial air. "Labor sells out the unions. The Liberals are conservative, and their coalition partners, the Nationals..."

"Let me guess: don't have much of a nation-wide base."

They high-fived. He grinned. "Got it in one. So you see Australian politics is pretty simple. Then we have the minor parties. The Democrats self-destructed because they couldn't choose a leader. One Nation stands for dividing the country racially and the Greens..."

"Don't tell me: the Greens are anti-vegetable."

Wicked grins were exchanged, followed by energetic snuggling. A coffee cup landed on the floor.

"Oops. Better not use your nice glasses in bed."

She laid a finger on his nose. "*Our* glasses."

"Better clean up the mess," he said, his nose tingling as her finger withdrew.

They were almost through with the papers, when the phone rang. It was Martin's dad. "When are you two showing up?"

"It's still pretty early dad... Saturday..."

"I wake up early and anyway, what are you waiting for? I want to meet your lovely lady."

"OK, dad. We'll get the next train south. Check the timetable – Ipswich line Brunswick Street to Graceville – if you want to meet us. Only takes us ten minutes to get to the train." He explained how to find the Transinfo web site.

As he hung up, Angie asked, "Lady?"

"Sorry, my dad is kind of old fashioned. I've been working on the twentieth century thing."

"Twenty-first."

"I can count. Catching up two centuries at once is a big leap. Let me check the train times."

The next train was in twenty minutes, so they set out straight away.

On the way, she asked, "So if he is so out of touch, why did you think my plan would work?"

"I don't know. The twenty-first century is still young. Maybe the leap won't be so great from where he is now."

As they neared the station, and Chinese text started to appear on street names and shops, she said, "You didn't tell me you lived near Chinatown."

"You walked through it last night."

"Oh. What else have I forgotten?"

"Nothing important – you just offered to give up your PhD idea and stay at home making lots of babies."

"I did *not*."

"You *did*. But I insisted that if it had to be one of us, it had to be me."

She cuffed him.

"Lucky you're smaller than me. We have laws in this country against child abuse." He rubbed his shoulder in mock pain.

It was a short wait for the train.

"Look at that graffitti." The train was sprayed with various crude, mostly indecipherable slogans.

"Yeah. Street art around here is not that creative."

"No need to apologize – we have some crap graffitti around Berkeley as well."

The train stopped at the next station, Central, with no urgency about moving on again. She looked around.

"Trains all stop here for a few minutes to allow for changing trains. Not much to see here – the platforms are below the main station level," he explained apologetically.

Their train went through several more stations: the last big one on the route, Roma Street, then Milton, Auchenflower... Martin explained local attractions at each.

"Indoor... what?" she asked a few stations later.

"Indooroopilly." He repeated it a few times.

"What kind of name is that?"

"I don't know. You live near a place all your life and never think. If it's not named after somewhere settlers came from or a colonial personage, it's probably Aboriginal."

"Aboriginal? What sort of language is Aboriginal?"

He laughed. "Not one, many. Dozens, I don't know. Some groups refuse to talk to each other. It is really complicated – and I know almost nothing."

"Let's find out together, then."

"Another project." He put a hand on her shoulder.

The train arrived at the Graceville station.

"There's your dad –" Angie pointed.

"How did you know?"

"First, I saw him on Skype. Second, no one else is looking so excited to see the train – except maybe that two-year-old who looks like he's never seen one before." She pointed at an excited toddler on the platform, trying to escape from his mother. "I think the first one is dad."

She gave him a dark stare. "I'm really glad it's not the other one. Or this has to be something out of *Doctor Who*. Let's go. I want to meet him."

It was a slightly hesitant meeting at first. If shyness was hereditary, Truscott senior was obviously the source of Martin's reticence with strangers. She tried to break the ice. "What would you like me to call you? Dad? Your first name?"

He thought for a bit, as they started walking towards the house. "You know, hardly anyone calls me by my first name. But your choice. George or dad. What the hell. Call me whatever you like." He had the pleasure of her impish grin for the first time. "Now wait a minute. That isn't an open invitation. Something nice."

"How about dad, except every now and then, to remind people you have a first name, George."

"Fine with me."

A black bird, about the size of a large chicken, scooted across the road right in front of them, neck extended, tail in a vertical fan.

"What's that? Some kind of turkey?" Angie stopped and stared as it vanished under a hedge.

"Scrub turkey," Martin explained. "I don't know if it's really a kind of turkey. We have trees with names like something oak, that look nothing like an oak."

"Oh. So do you have something like thanksgiving here, I mean, a turkey day?"

"Not really – some people eat them for Christmas but many others have seafood."

George touched Martin on the shoulder, as they started walking again. "You know, son, I bet I know one thing about cooking you don't – how to cook scrub turkey."

"Oh? Dad, I thought you've never cooked anything more complex than a sausage on the barbie."

His dad gave him a serious "you don't know everything about your old man" look, then went on. "You get a large pot of water. Put in a rock. Then the bird. Boil it till the rock's soft enough to eat then throw away the bird and eat the rock."

They all laughed and Angie clapped him on the back. "Marty picked up some of his sense of humor from you."

George smiled. "By the way, you may notice I'm not here in my car. Carbon guilt. You're starting to get at me." Martin and Angie exchanged charged glances.

"Gee, I thought you were just exercising more – working off that cholesterol problem," Martin said.

"Well, maybe I should do that too – how's that swimming going?"

Martin looked at Angie. "Good point – we should go for a swim at UQ. Should have brought our things."

"I was going to say you could borrow something from me, but of course I don't have girl's togs."

"Never mind, dad." Strange to hear her calling my dad that. Martin smiled as she continued. "I'm still a bit tired. If we don't get around to it, maybe tomorrow."

They reached the house. Angie stopped to survey it. "My, this *is* nice."

"Wait till you see inside. My dad's...uh, *dad's* done a great job of fixing it up."

George showed her around. The new kitchen, the bathroom, the ensuite bathroom attached to his bedroom, Martin's old room, a guest room, his study. "Like a lot of these houses, this one has had some add-ons. It used to only be three bedrooms, bathroom, kitchen. An owner or two before me added the ensuite bathroom by stealing a bit of the wrap-around verandah. Maybe the same person, I'm not sure the order of alterations, took over the verandah on the other side to add in another bedroom, which I use for my study. Now the old verandah looks more like a front and back deck – more modern, but not as good for breezes."

"And nice you've split off the garage from the store room and laundry downstairs," added Martin.

"Great kitchen," said Angie, obviously forgetting the scrub turkey conversation. "Cooking must be a Truscott thing."

"Not really. I did it up because I was told it added value to the place. About all I make in there is tea. Speaking of which, can I make everyone some tea?" He looked at Angie, then Martin.

"Why not? Angie, do you like tea?"

"I don't get it often at home, but I'll give it a try."

As George made the tea, she had a look at the yard from the back deck. "Is that your mango tree?"

"My mango tree?"

"The one where you used to talk to birds."

"Yes. Yes, of course. I just hadn't thought of anything here as mine for so long."

"Let's go down there. Are they smaller than scrub turkeys?"

"A lot smaller – dad, I'm showing Angela the back yard."

"Tea will be ready in a minute."

He took her down the back stairs, and showed her the spot under the tree, where you couldn't see the house. The shrub screening them from the house must have grown – it seemed the same size to him as it did when he was small. He must have stopped going there around about the time his mother died. He shared all this with Angie.

"Then you started on the games?"

He nodded. "Seriously. Compared with before."

It was a solemn moment. They sat there in silence for a few seconds, and she pointed slowly. A little black bird with a long tail had landed and started to contemplate them, soon followed by another. They watched for a few minutes, until a voice floated down from the house: "Tea!"

"Magic place," she said, as they headed back for the house. "I hope to spend more time with your little friends."

"Or their grandchildren. I don't know how long these little guys live."

Back inside, tea was passed around. Martin had his black, no sugar. Angie had a taste, decided it was OK, but decided to try hers with milk. George passed around a bag of Tim Tams.

"Tim Tams?" Angie enquired, reading the name off the label. She took one. It needed no further explanation – chocolate shell over two crunchy layers, with a soft chocolate layer in the middle. "How long exactly is it going to take to get Australian citizenship?"

They all laughed.

That reminded Martin: "Dad, on the way here, we were wondering how Indooroopilly got its name."

"You know, Martin, I'm not a hundred percent sure, but I think it's an Aboriginal name."

"Well, when you go into politics, you'll need to acknowledge the traditional owners of the land every time you make a speech, so you might as well find out."

"That's an interesting custom," said Angie. "Do all the politicians do that?"

"I don't know. Only the ones I've heard speak. Labor, Green, Democrat."

"Wait a minute, you've lost me there. Did someone say I was going into politics?"

"Yup, dad. The Greens are trying to persuade me to nominate as a lower house candidate, but I think they have the wrong Truscott. It should be you."

"Me? I don't know anything about politics. I vote Liberal and that was enough to freak out my bush relatives."

"Bush?" Angela asked.

"Everything outside the cities pretty much. That's where most of the Nationals vote comes from," Martin explained.

"But, dad you told me you don't talk to your relatives anymore. What do you care about what they think?"

"Yeah, but Greens. Aren't they rather, ah, left, kind of socialist?"

"Not really – not so much anyway. There's a bunch of other greenies way out to the left." He explained about Julie's friends. "The Greens start from the environment then go to social justice. If you actually look at a lot of their policies, the more libertarian side of the Liberals could be at home with them – gay rights, for example."

"Oh, yes. I suppose I've learnt some lessons on that front."

"And then, dad – the Greens really want me for credibility on climate change. But who would be more credible than an engineer who's delivered huge projects, and has voted all his life for the Liberals? If you changed, so could anyone else."

"Martin, you make me sound like the last dinosaur."

"No, George. The one that evolved." Angela put a hand on his shoulder.

He laughed. "You make this sound so appealing."

"Dad, I would do it except for two things. The PhD project I want to do is really important." He explained about the West Antarctic Ice Sheet, and the need to model it accurately.

His dad nodded. "I thought you would make a brilliant engineer. That's a problem on a vastly bigger scale than anything I've ever seen. And the other thing?"

"I have a young wife – officially in a couple of weeks, but we've been married in effect since I met her in Boston. I miss her terribly every time we're apart. Is it very selfish to want to have some time with her?" His dad glanced at Angie, but Martin kept his eyes locked on his dad.

His dad softened. "No. No, son it isn't. Not if there's a job to do but someone else could do it."

"You mean you're in?"

"I really don't know if I can do this. I'm not even a member."

"We can help with all that," Martin said. Some forms appeared as if by magic.

"And I don't know if they'll want me."

"I can't imagine why not," Angie said. "Anyway, I assume they will go through some sort of screening, I mean they don't just take the first person who signs the nomination form, do they?"

Martin shrugged. "I know as much as you. I've only been a member for a few days. Do it, dad. It will at least give those hillbilly relatives of yours a wake-up call."

George smiled. "I can just see them now, turning their noses up.

"OK. Let's give it a try, but you will have to give me a lot of coaching. I have no clue what a press conference looks like, let alone how to fend off the media sharks. I hardly know anything about climate change, and social justice – I'm not even sure I know what that is."

Martin looked him straight in the eye again, this time holding his gaze. "Dad, it's about mum living her life the way she had to without some hillbilly relatives tearing her down."

Angie meanwhile was starting to wilt. Martin, noticing this, changed the subject. "Dad, we can leave the stuff here. Go to the Greens web site if you want to find out more. I'd like to take Angie home to show her my neighbourhood while she's still compos mentis."

Once they were on the train, Angie said severely, "You may think I am not totally in command of my faculties, but Martin My Husband Since Boston Truscott, I hope you never accuse anyone else of manipulating you again."

"What did *I* do?"

"You hit just about every one of your dad's buttons. Did you see the look on his face when you accused him of tearing you away from the one you loved?"

"I didn't exactly..."

"No. Not exactly. But you know how hard he took it when your mom left him. I know exactly what you mean because I hate it too when we're apart, but that was the wrong thing to say."

He nodded slowly. "I really didn't mean it to hit him like that. I just repeated what I said to the Greens delegation. I should have thought. You're the biologist. What animal is lower than a snake's anus?"

"We've been through that already. We know you're a great ape with a racoon juvenile stage – or was that a homo sapiens?"

"Whatever." He felt better – but he fumbled for his phone. "Hi, dad."

"So soon, Martin? Do you want me to run for something else?"

"No, dad. I just want you to know, I want you to feel right about this if you go for it – not because I pushed you into it."

"Martin, don't worry about me. I had a kid growing up in my house who played head games with me non-stop. No one in our family is manipulating anyone anymore."

"I just wanted to say I was sorry, dad. I kind of feel I was insensitive in some of the things I said."

"Tell Angela thanks."

"For what?"

"She knows."

After the call ended, he asked her, "What's up? Telepathy?"

"No. Your dad and I exchanged a look when you did that guilt thing. I did a throat slitting mime in your direction. I think we covered that up rather well, didn't we?"

"What species did we decide *you* were?"

As he pointed out landmarks on the way, she said, "You know, maybe a swim *would* wake me up. Should we see how I feel after I unpack my stuff? How long does it take to get to the pool from home?"

"Depends how energetic you are. A couple of kays walk to a bus that runs every ten minutes to the uni, or two buses from closer to home – could take any amount of time. The bus timetable is a joke."

"OK – let's see how I feel. The walk could be good anyway, to get a feel for the layout."

"Yeah, Brisbane is designed to fool the space aliens. No real grid pattern; the river gets in the way."

"What was that last stop? The train went under a big high rise."

"Toowong."

"Another Aboriginal name?"

"I guess. There's a shopping mall above the station."

"Convenient. Do you ever go there?"

"Not really, not my part of town."

"How about we get lunch stuff there tomorrow and cook for your dad? Convenient, like I said, right on the way."

"As penance for a few decades of head games?"

"No, silly boy. To put that fine kitchen of his to good use."

The train wound its way through increasingly urban stations, until they hit Brunswick Street. "Our stop." She put her arm around his shoulders as they left the train.

On the way home, she was paying more attention than before. "Still look pretty awake to me," he observed.

"Not too bad," she said. "How about I unpack a few essentials, we go for the swim, then home to finish up? It'd be a pity to waste such a nice day. Is this average for summer?"

"It gets more humid later in summer, and it rains a fair amount. But – yes – let's not waste too much time before swimming."

Back in the apartment, she faced her as yet mostly packed luggage. "Let's see. I think my swimsuit is in this one, and here's another essential. An old family favourite. I told them to order another one from Amazon. This one has good memories." She pulled out a broad, well-read paperback and passed it to him.

The Silver Palate Cookbook. He turned it over, read the story behind it on the back, and noticed there was a bookmark in it. At that page was a recipe for lime mousse. "Hey! I know this!"

She smiled. "Like I said, happy memories. That was the second time I made that recipe – from memory cause I didn't have the book in Boston – here, put it with the Riedel glasses. They belong together."

She kept rooting through her clothes, putting things aside.

"I'll get you a towel in the meantime." He went to fetch one. "And while you get your stuff together, I'll check if dad is OK for lunch tomorrow." He was.

They walked through the city centre, Angie admiring the older buildings juxtaposed with some imaginative new high rises. "Some nice buildings – I like the mix of old and new."

"Yup. Not always as it seems. Some that look old at street level are just facades. Maintaining the heritage look, I think they call it."

As they neared the river, he pointed out an ornate building. "I think that used to be a government building. It's a casino now. The river is just a block or two that way, but our bus stop is before the river."

"I wanted to see the river."

"Don't worry, we cross it a couple of times."

True to his word, the bus crossed the river soon after starting. She looked out over the water. "Ooh! What's that?" She pointed at a big blue catamaran with a glassed-in cabin and open-air seating at the back and front, full of people with a festive look to them.

"City Cat. Ferry. I don't usually use it because buses and trains are quicker most places I go. Popular with tourists. We could take it on the way back if you like."

She nodded.

He pointed out landmarks on the way for a while, but she leant on him and said, "Next time. Just keep me awake." At the university, he pulled her off the bus. "Let's take the scenic route – only a couple of minutes longer."

They walked past two large ponds full of water birds, then

along a long curving path up a grassy hill. Taking in the sights was keeping her awake. "This is beautiful. Is UNSW anything like this?"

"Not quite as nice. No ducks. But Sydney has its attractions."

"Opera house."

"Yup – and beaches much closer to the city."

They reached the pool. Jet lagged or no, Angie was still easily the better swimmer.

The City Cat trip on the way back may have taken longer than the bus but, standing on the front deck with wind ripping through her hair, she looked more awake – and responded more to his landmark narration.

*   *   *   *   *

The next morning, they made a lazy start. Angela claimed she was almost in the time zone. After a slow breakfast, checking emails and reading papers, they started contemplating the expedition to Graceville.

'What should we make?" Martin was still feeling lazy.

"I don't know what's good here this time of year: if you don't have ideas, let's check out the shops."

"Dessert?"

"Oh, yes, I *do* have an idea for that."

He grinned and nodded, glancing at her cook book next to the wine glasses.

"How about we bring our swimming things this time?" Martin retrieved them from the bathroom. They were still slightly damp. He found a couple of old shopping bags to pack them in. Angie found a small backpack in her luggage. "I've also got one –" Martin found his – "we should be able to fit the shopping in these."

Angie was looking puzzled. "I was thinking of something I needed, something for the kitchen."

"Never mind. Maybe it'll come back when we do the shopping."

Soon, they were in the train to Toowong.

At the mall, Angela looked around. "There's a fruit and vegetable place there."

"And a supermarket right next door." He pointed out the Coles store.

"But there's that something I'm missing. Is there a kitchen store here?"

"I don't know, let's look around. What is it? Remembered yet?"

"Don't know. I'm sure I was thinking of it when we were planning the trip. Maybe it'll come back."

Martin was getting impatient when they hit the third shop and she still couldn't remember. Then she stopped. "Yes!"

She was looking at a pasta machine. "I've always wanted one of these. I was visting Dale's home. . . " she saw the look on his face. "You're forgetting he's gay."

"Sorry. Tell me about past boy friends some time. I told you my entire history."

She nodded. "Julie. Mine's a longer story, if less interesting. Anyway, his mom made this great pasta – eggs, semolina. I lusted after the machine but, you know, something to get when I had my own place." She smiled at him.

"Well, you got our extravagantly expensive wine glasses. Let me get this."

"No, no. It was my idea. I should get it."

A shop assistant approached, cheerfully garbed in a black apron. "Can I help?" She sized up the situation immediately. "How about you get one for each other?" She smiled engagingly.

They both laughed. "Compromise." Martin picked up an extra attachment for making filled pasta, big triangular ravioloni. "I'll get this."

They headed back to the food stores with the pasta maker in two large bags.

"What should we fill them with?" he asked.

"I don't know. It was your idea to get that part."

"Sulking?"

"No, jet lagged." Angie gave him a tired look.

"OK, so that's what it's called now. I'll remember that for when I need it. Ouch." She nudged him in the ribs. "OK, how about a mix of spinach and ricotta –" he tried to remember an Italian cooking experience from the distant past – "with a hint of nutmeg?"

"Sounds good. Should I get the spinach here," she indicated the fruit and vegetable place, "and meet you inside, since you're so slow?"

He mumbled something that sounded like "Who's my swimming coach then?" and went into Coles.

They met at the deli section. She showed him her purchases – a bunch of spinach and half a dozen limes.

"Great – I can guess what the limes are for, but I need some help. How much ricotta to get. Wouldn't you say about a hundred grams a person is about right?"

"No clue – I was brought up in ounces and things, sorry." She leaned on him, starting to look tired again.

He asked to see a piece weighing about three hundred grams, and they decided it looked a bit much, but he nodded. "OK, that will do – better too much than too little." He showed her his other purchases: a dozen eggs, a small bag of semolina. "Let's see, now... parmesan. Reggiano only appeared in this country a few years ago – raw milk cheese was supposed to be a health risk."

She raised her eyebrows. "To think, we Americans have been testing this stuff for you for decades."

"And the rest of the world, for how long?" He paused to contemplate. "What else? Cream for the sauce, nutmeg." He found whole nutmeg, and she went scooting off to find cream, and returned with a carton and some butter.

"Butter and cream also needed for the mousse, and we don't need all those eggs just for the pasta, so it can share those with the dessert course."

As they headed for the checkout, Martin said, "I'm such an idiot."

"Devolving to a great ape again? Not back to raccoon, I hope."

"No, well, maybe ape. I forgot dad is watching his cholesterol."

"Hmm. So I suppose we need some alternative to cream sauce. What was that thing you made in Boston? That was pretty good."

"Wouldn't quite work, unless we give up on the filled pasta. Let's get some nuts and olive oil, though. A lot of the commercial stuff has no egg, so it must be possible to make pasta without egg."

"Right – we can work some of this stuff out at the house."

Martin found some macadamias in the bulk section, and Angie called him over to the other side of the bulk containers, as he finished measuring them out. "How about some of these?"

"Pine nuts? Why not? I've only used them before to make pesto. And speaking of pesto, let's get some garlic too – I think dad said something about garlic being good for cholesterol."

"You make pesto? Better than straight out of a bottle?" She feigned surprise.

"Of course. But next time. We want to use the new toy."

They found a small bottle of olive oil that didn't look too industrial, then headed again to the checkout.

When they arrived at the house, George Truscott was waiting impatiently. "I was expecting you to get here half an hour ago." He focused on the copious shopping bags. "Good grief. Did you buy the entire store?"

They were both staggering under the load. The pasta machine plus extra accessory turned out to be rather heavy after carrying them a few blocks. Martin dumped them on the kitchen table. "Dad, I need to check something – pasta ingredients."

As Martin walked to the study, his dad said, "Of course, Internet. I was going to say we have no cooking books in this house."

Martin reappeared in a couple of minutes with a piece of paper. "Let's try this – semolina flour. I guess that's the same as the stuff we bought. Warm water. And maybe a splash of olive oil. Lots of kneading."

"That would be to work the gluten." Angie was at the fridge, packing away the cold stuff.

"Thanks, Ange – I think I showed you how to make pizza at some point."

"Kids, please – save the competing for the Olympics."

"OK, dad. Sorry. Ange, if I get the pasta dough going, can you wash the spinach? Three times is good to get the sand out, but save the water. For the garden."

"OK, boss. But then I want to start on the dessert. The surprise this time will be to serve it cold, the way it's supposed to be. Save it for an hour or so after lunch to chill." Meanwhile she was surveying the equipment, and selected a small saucepan and a stainless steel bowl to make a double boiler. "I *knew* there was something I wanted from the kitchen store for doing lunch."

"Not the pasta maker?" Martin's dad asked.

"George, that was an unfulfilled craving. I actually needed a double boiler for what I was planning from the start. This will work though. I've had to improvise this way before." She took the spinach over to the sink.

George Truscott surveyed the scene, feeling out of his depth. "Anything I can do?"

"Put on some music, dad."

"OK, Martin. How about some wine? I have a nice Tyrrell chardonnay, designed to age, and I've kept it for ten years." He turned towards the fridge. Martin and Angie, at work at opposite ends of the kitchen table exchanged meaningful glances, causing him to pause. "What was that about?"

"Dad, you have to admit, your heart is to the left of the Liberals." He explained the latte and chardonnay thing.

"Hmph. I bet those clowns in Canberra don't drink beer straight from the bottle and drive around in utes."

Angie and Martin nodded; he added: "Exactly what we thought, dad."

George produced the bottle. "Still a bit dusty from the cellar, brought it up last night."

"Should be good." Martin went back to kneading his dough. "Looks good, smooth, silky." He looked up. "The dough, I mean."

His dad laughed, as he put the bottle away. "So what exactly are you making with that?"

"The pasta machine has this attachment for making filled pasta – big triangles."

"Right, so what's the filling?"

"Getting there. It says here –" he held up his of piece of paper – "it should rest twenty minutes after kneading, so I'll do the filling then. Spinach and ricotta – not too much ricotta, it's low fat as cheese goes. We remembered your cholesterol."

"I was beginning to wonder when I saw that butter and cream."

"Are nuts and olive oil OK for you?"

"No problem. The doc says they're good for cholesterol."

"Great. We'll make two sauces then. You'll have to skip dessert." His dad looked so disappointed, he softened: "OK, a small taste then."

The pasta looked ready to rest so Martin unwrapped the ricotta, and rescued the spinach from the sink, where Angie had not only cleaned it, but trimmed off the roots. She was meanwhile whisking over the stove.

Martin searched for salt. Eventually, he found a stash of old airline meal sachets. There was just enough. The spinach cooked quickly, under the influence of the powerful gas burners. In the meantime, Martin found a chopping board and a small knife, and started attacking the nutmeg, shaving it. Ange said, "Doesn't your dad have a grater?"

"I'm sure there's one somewhere, but this works better. Learnt this trick when I didn't have a grater."

He retrieved the spinach from the stove, drained it, poured cold water over, then squeezed the moisture out. "Ouch. Damn spinach. Hot." He chopped it vigourously in retaliation.

"You want to be a big man and take on vegetables that can fight back."

"Whose side are you on anyway? Take a look here." He had the filling assembled. "Taste."

"Wow. I didn't know spinach could taste like that."

"Same trick as in Boston: salt to make it cook hot, wash it out afterwards."

She tasted again. "Maybe a touch of pepper?"

He produced some airline pepper sachets. "Speciality of the house."

"Exotic. Meanwhile mine has turned to custard, so let's put it in to chill, and focus on the pasta machine – if it's had its twenty minutes."

George returned, some Chopin playing in the background. "I hope you kids don't mind some classics for a change."

They turned to the pasta machine. It needed a little assembly but that side was pretty obvious. Martin clamped it to the kitchen table and flattened out the instructions on the English page. "OK, so we need to put the dough through on the widest setting, then keep doing it on finer settings, until we get down to setting five. Then we put the other attachment on, and make filled pasta."

Angie nodded, "I have seen this done before."

"Sorry, I forgot. So tolerant, letting me play with your toy."

"Sounds easy," George said, cutting into the hint of building competition. "Not that I know anything."

Martin divided off about a quarter of the dough, and mangled it through the machine. It came through... mangled. It had holes in it, and stuck in various places. With some effort, he managed to get the whole sheet through.

"What are we doing wrong? The dough felt right – firm, not too sticky."

Angie picked up a piece and puzzled over it. "Pretty much as I remember it."

"If that was a machine," George offered, "I'd say it was short on lubrication."

"Dad, you're a genius."

"Really, I'll just go down to the garage."

"No, no. Not machine oil. What do you use to stop dough from sticking?"

"Flour?" Angie asked.

"Exactly. You did pay attention to my pizza making."

"Huh. Knew that already. George, do you have any flour?"

"Well, no. Don't need it for making tea or – scrub turkey. Shops aren't too far away."

Angie grabbed a mug and thrust it at him. "Is the concept of getting a cup of flour from your neighbour known here?"

"I don't know – I never even talk to them. Not the best on sociability." He looked awkward.

"Here's your opportunity. Go! We're getting hungry here." Angie pushed him towards the front door.

George reappeared in a couple of minutes, an energetic-looking woman, lightly greying, hustling him back, with the mug now full of flour.

"Well, George. Pleased to meet you. After twenty years living next door. Just as well you eventually needed some flour. Well, don't just stand there. Who's everyone else? Muriel. Muriel Spark. Mad Muriel in the trade."

Martin completed the introductions. "I'm Martin, George's son, and this is my partner-to-be-wife, Angela."

Muriel stopped in front of Martin. "My, my. Didn't I see you on TV chewing apart some clown from Oxford or some such?"

"Guilty as charged. Harvard."

She surveyed him at arm's length, a hand on each shoulder to measure the distance. "To think, that wretched child who used to annoy all the neighbours would get that far. Well done, boy."

Martin crumpled. "Didn't *anyone* like me?"

"I remember a boy who didn't like himself much," Muriel said.

"Gosh. Are you always so direct?"

Angie laughed. "Marty, look at you. First, your life is ruined because everyone lies to you, now you can't take straight talk."

Muriel smiled. "How else would you like me to be?"

Martin looked her up and down. "I don't know you. But, uh, what do I have to do to have you on my side?"

She laughed. "Just be yourself. I find that works."

George said, "What did you mean, 'in the trade'?"

"Real estate. Been an agent for twenty years. When I started, they all said with my attitude you'll never last a week. Got to learn the code. If it's too small, 'cozy'. If it's dilapidated, 'opportunity'. All weasel words. I just tell it like it is. 'Won't last a week.' Huh. I made enough money in my first month to buy this house –" she pointed next door – "and I've been the top seller in the western suburbs every month since, even when I take holidays."

"Oh," said George. "And what do all those people say now?"

"No idea. They all threw in the towel years ago."

"Dad, tell her about the idea of selling the house."

"What idea?" demanded Angie.

George explained about the 1893 flood levels, as related to the multi-metre sea level rise threat.

Muriel said, "Are you serious?"

"Ask him." George pointed at Martin.

Martin explained the NASA and British concerns about the West Antarctic.

"So you're saying for business as usual, whatever that means, not only this area but half a billion homes are going to be flooded, and you want to sell?"

"Well, yes..." Martin began, with a queasy feeling.

"Well, *no*. That's the most cowardly thing I ever heard. You know there's this problem. You have the public's attention. If you can't bloody well get something done, who can?"

"Martin, dad –" Angela looked at each in turn – "she's right. If George runs for the Greens, and we campaign hard, there's a chance we can get the message through – at least in Australia."

"Good girl." Muriel took her by the hand. "Running for Greens, eh? Where, here?"

"Actually," Martin explained, "they asked me to nominate, and I live in the city. I guess dad should run somewhere closer to home if they choose him."

"Very good. You need to explain all this stuff to me in more detail. The press is so confusing. Meanwhile, here's your flour." Muriel pointed at the mug George was still brandishing.

They all begged her to stay for lunch. "I thought you'd never ask. What are you making?"

Martin pointed at the pasta machine, and explained the problem. He laid a lump of pasta out, dusted it with flour and – "Perfect!"

"Let me try!" Angela took charge for a while, then they alternated.

A few more runs through the machine, and they had four sheets of pasta, ready to be trimmed and filled. The filling attachment proved to be a minor challenge. After a few cranks, Angela stopped Martin. "It's eating its tail!" She untangled the end that was going back up the machine.

Eventually they had four neat rows of triangular pasta,

and some off-cuts laid out flat.

"Right, now – sauce and cooking the pasta." Martin looked for a couple of smaller pans for the sauce, and something to serve as a large pasta pot.

"Muriel, dad has a cholesterol problem, so we're making a sauce for him with olive oil, macadamias, garlic and pine nuts, and a cream sauce for the rest of us. Which do you want?"

"Both sound good."

"Both?"

As the water was boiling up, Angela took charge of the cream sauce, while Martin chopped nuts. With the sauces cooking, Martin tasted the nut one, and offered some to Angela. "What do you think? Splash of wine?" she nodded. The sauce sizzled its objections, then gave in and slurped up the wine.

The water started to boil, so Martin tossed the pasta in, including the off-cuts.

"How long should they cook?" Angela asked. "Dale's mom made unfilled pasta and I remember that as cooking pretty fast."

"Until they give up."

"What?"

Martin explained a TV cooking show he'd seen in which an Italian matriarch had explained how to cook gnocchi. "Put them in the boiling water, and cook them until they give up."

Everyone thought this was pretty funny. In a couple of minutes, the filled pasta floated to the surface. "See: they're giving up! Taste test." Martin took one out of the water, and tossed it around to cool it off.

"Done!"

Angela said, "Let's taste Muriel's choice – the combined sauce."

"Good plan." Martin found a couple of spoons, and mixed the two sauces on one of the spoons.

Angie watched his expression. "Gimme, gimme!"

She tasted, then: "Muriel, I think a few more of us will go with your idea."

Everything was pretty much together.

George surveyed the kitchen. "You know, it would be silly to eat in here. First, we'd need to clean up. Second, it's really nice outside. Third, it's time the back deck was introduced to something more sophisticated than a barbie."

Angie agreed. "Good idea. I have some final assembly to do on the dessert now it's cooled off a bit. I'll do that while you set up outside." Muriel watched in fascination as Angie juiced the limes and started whipping cream. "Shoo! You're a guest. Tell them to get started. I'll be out in a minute."

Muriel went out, and helped with the final arrangement of the back deck table. After taking her seat, she took a taste of the pasta, then of the wine. "Mmm. That dreadful boy has become a real treasure, if only for his part in this meal." She looked at Angie, as she emerged from the house. "Well done, my girl. You could make a fortune as a counsellor."

"Me? I did nothing. I was just lucky to be there when he was good and ready for me. Everyone else did the hard work."

Muriel smiled. Direct, or no – there was a time to keep quiet.

*   *   *   *   *

"That was *interesting*." Angie looked back at the house.

"Wasn't it just. I never knew we had a neighbour like that. Not even a hint."

"I think raccoons are nocturnal." She gave him her most serious "trust me I'm a biologist" look.

"You'll never let me forget that."

"Of course not. But the interesting thing is they're still talking. Look at them." Angie looked back at the house. George and Muriel were in animated conversation at the front door. "They look like new best friends. I thought we'd never get away, but once they got going, they only seem interested in talking to each other."

"Let's stop dawdling then – leave them to it." Martin reluctantly turned away from the house.

They scurried towards the station, the pasta machinery back in its boxes and shopping bags, their swimming things in their backpacks. Martin had the larger box, Angie the extra attachment.

While waiting, Martin mulled over options. "I didn't expect to have so much stuff with us when we left home, but we can dump the pasta stuff in a locker while we swim. Too much effort to go home and back, when the pool's between here and home."

"Does the train go near the uni?" She was learning the lingo.

"No, but if we get off at Indooroopilly, there's a big bus stop at the shopping mall. We should be able to get a bus there."

"On the same ticket?"

"Yes. A ticket works on all transport modes in the zones you paid for."

"Cool."

The walk from the Indooroopilly station to the shopping mall, encumbered by heavy pasta machine parts, was rendered longer and more uphill than Martin remembered, and the buses were right at the top of the hill, behind the building. "Lucky we're doing this diversion to get some exercise, cause we're getting some right now."

"Oh, stop grumbling. Oops, I forgot – you're older. Can I carry the heavier part for a bit?"

He pushed her hand away. "Damn reverse psychology."

At the bus stop, an electronic sign announced when the next few buses were due.

"Very high-tech," she observed.

"Very. The signs are linked to an AI creative writing system."

"Huh?"

"Look at the bus numbers, and times – now watch for what happens."

After a few minutes, she said, "Is this totally random?"

"I'm not sure. I think they slip in a correct one every now and then to confuse the space aliens."

"Oh, right – like the confusing street layout. We're going to have to fix this space alien prejudice thing. Add it to your dad's platform?"

A suitable bus finally showed up, and they boarded along with an assortment of people from skinny-looking people of student age, to shoppers who looked big enough to put the bus off balance. "Supersize me," Martin whispered to Angie.

"You not me," she whispered back. The bus driver didn't know what to make of the wicked grins, and just grunted as they showed their tickets.

Martin led the way past the driver.

The bus stopped on almost the far side of the campus from the pool. By the time they got there, Angie had seen another side of the campus, and Martin's arm was stiff from carrying his part of the machine. Once the shopping bags were in the locker, he massaged his arm.

She was unmoved. "No excuses! I expect you to be swimming better after all that practice. Or did you forget everything your coach told you?"

They went back to the same routine – Martin going first, a critique at the end of two laps, Angie then doing two to demonstrate, and sending him off on his own. After a few lengths like this, Martin stopped.

"Tired already?" She stopped next to him.

"No – look at that sky." Thick black clouds were rolling in.

"Rain?"

"Worse. Look how bright green the glass is looking."

"Radiation?"

He splashed her. "Maybe worse – hail."

She did a mock salute. "Hail my caesar!"

He splashed her again. "Let's go. Quickly, before we're caught in it."

They rushed into their respective change rooms, and shortly after started rushing towards the bus stop – the one on

this side of the campus. "No, wait – we forgot – the locker."
Martin went back and retrieved the pasta equipment. They
made it to the bus with a minute to spare. As the bus was
heading in to the stop, big, fat rain drops started plopping
down.

"Now I understand that sign on the bus." Angie pointed
to a sign on the front of the bus: *Please hail driver.* "We're
supposed to wait until the hail starts then throw some at the
driver."

"I think that will have the same result as that thing you
said to the nice immigration people at Sydney." He was balancing the big pasta machine box on his knees, and patted it.
"Lucky we didn't forget this...oh, crap – do you know what
we did forget? Dessert."

"Uh, oh. And your dad is watching his cholesterol."

Martin pulled out his phone. It rang six times. "Dad – I
hope you didn't eat all that dessert."

"I thought if you forgot it you didn't want it."

"Cholesterol..." Martin interjected.

"I know, I know. There's a little left."

"He didn't eat it alone!" There was a voice in the background.

"Dad, is that Muriel? Still there?" Angie winked at him
as he asked. He wasn't sure how to take this development –
things were moving fast on a front where he'd never known
any action.

"Um, Martin, if I am to be a candidate, I'm going to need
a campaign manager who knows a bit about people. Your
name being so good as a start, you know, but..."

Muriel grabbed the phone. "What d'you mean *about* people? I know people! No one in the western suburbs didn't either buy their house from Muriel Spark, or have a story about someone who did."

Off to the side, George was heard protesting, "I didn't..."

"Of course you bloody well didn't – I mean real people who have a life."

When Martin pocketed his phone, it felt as if it should still be quivering.

In the city, rain was starting to come down seriously, so they dashed for cover. "This is Queen Street. There's a pedestrian mall this end of it." Martin pointed at the start of the pedestrian mall.

"And all of it's wet," she pointed out. "Are we going to wait here or make a dash for it and get wet?"

"Let's dash."

Despite taking advantage of whatever cover there was, they were drenched, and the pasta maker boxes were looking soggy where the shopping bags hadn't protected them. Martin dumped them down, and ripped his shirt off. "Let's toss these in the bathroom." As their clothes came off, the hail thundered down.

"Protect me," she yelled in mock terror. "Mmmm," she added as he complied. "I should be a coward more often."

19   Final Take

THEY WERE DRIED OFF and packing stuff away. "Maybe we should have left the pasta machine at your dad's – there's no way we have space here to use it," Angela mused.

"But think of all the fun we had sprinting with it in the rain – a new olympic sport."

The phone rang. Martin picked it up.

"Yes, this is the Martin Truscott who has a father called George . . . Mabel who? . . . Great aunt? No, my dad didn't tell me he had an Aunt Mabel in Mount Isa . . . John who? My dad's brother?

"Don't you know what happened? . . . Yes, I suppose you *are* in a position to . . .

"Why don't you tell him yourself? . . .

"And why doesn't that apply equally to me?"

He slammed the phone down.

"What was that about?"

"It seems my dad has an Aunt Mabel in Mount Isa – a dump of a mining town in the far north of the state – that I never heard of, and she tracks me down to tell me something because she's too shit-scared to talk to my dad." Martin was shaking.

Angie touched him and he didn't respond. "Why did she call you? What was that about your dad's brother?"

"Seems he's dying of cancer, and she's too scared to talk to my dad herself."

"Because of what happened."

"You are such a genius."

"Hey, wait. What did I do?"

He shook his head as if trying to clear it.

"Those damn people ... trashed my mother, made me grow up without a family ... my dad was such a misery and I was a total brat. They destroyed my childhood ... and now they want me to help them patch up with my dad."

"But Marty, it *is* his brother. Maybe he would want to know ... "

Martin stood up abruptly. "You just don't get it." He fled to the door, wrenched it open and rushed out slamming it behind him.

He moved through shadows of shadows of buildings, all familiar-unfamiliar, old friends masquerading as strangers, or was it the other way around? He crossed bridges. Sat on benches. Found a City Cat stop. Lay on the bench. Cried a puddle through the slats. A drunk lifted his head, said tsk *through winey breath, right in his face. He staggered up. He crossed a bridge. He saw trees. They looked friendly. Then they did not. He saw familiar streets. He found himself going a way he'd been before. He found a familiar door. He knocked. A silhouette appeared against the frosted glass. Not real. Hinting at real. He'd been there once before. Twice.*

Returned a photo. Geraldine Simmons opened her front door and Martin almost fell onto her.

"Look at you. What happened?"

He blurted out the tale. Semi-coherent.

"Isn't your Angela here?"

He nodded indistinctly.

"Shouldn't I tell her where you are? What's your phone number? Does she have a mobile?"

He felt in his pockets in a blind panic. No phone. No keys. No wallet. Had he been robbed? He didn't know. He didn't know what was happening to him, where he'd been, how he got there. Other side of the river. How long? His mouth moved silently.

"Never mind that. Let's get you home." She found her keys, and hustled him to her car. "You do know where you live?"

"Valley."

"Give me directions."

She fastened his seat belt. The solid clunk made him feel strangely secure.

They drove in silence but for his occasional grunted direction.

She parked outside his building and took guidance to the right door. It was locked. Before his panic could set in again, Angie opened it.

Double vision? No. Julie was there too.

"What in the world...?" He wasn't sure who asked.

Geraldine brushed past Angie. "Later. Let's get this boy to bed, and get him calmed down. Geraldine Simmons. Pleased to meet you all."

She laid him out gently on the bed, then called out, "Water!"

Angie brought a glass.

Geraldine took it. "Drink this!" Martin complied, and some colour started to return to his face.

"Now listen. I'm going to have a chat with your dad." Martin tried to sit up, and she pushed him down. "Oh no. You stay right there. You said you were working on his attitude towards me, and I think you've done quite enough. I'm leaving you in good hands here, OK? Just relax. Get the girls to give you a nice massage, and forget about this whole thing if you can."

He tried to talk.

"No talking. Just relax."

A few minutes later, she stood up. "I have a very good GP who can handle this sort of thing well." She wrote on a piece of paper and handed it to Angie. "I'm hoping the worst is over, but if not, give her a call."

Angie started to thank her, but Geraldine shushed her. "It's not over yet. I just hope I'm doing the right thing. The last time I talked to George, it was through a lawyer."

She drove to Graceville, mulling over her options. Nothing for it but to cut to the chase.

She found the house easily, despite not having been there for more than ten years. The car was there so George would be home.

Knocking on his door was one of the hardest things she'd done in a long time. Geraldine Simmons: you've been to hell and back both ways. If anyone can handle this, you can. She straightened up as the door opened.

"You."

"Yes, me."

"What are you doing here?"

"Your son is in bad shape. We need to talk?"

"Martin?"

"What other son do you have? Your aunt Mabel phoned him, and it sent him into a state."

"Mabel? You'd better come in I suppose."

They stood just inside the front door, and Geraldine said, "Look, George. Your son is in trouble. We need to forget our differences and sort this out."

"We should sit down, I suppose." He led the way, and they sat stiffly.

"George, he arrived at my house barely coherent. The thing I managed to piece together is that Mabel phoned, and hoped he could pass on to you that your brother John is very ill. I imagine there are some hard feelings between you, and she thought he would be easier to talk to. He's been under a lot of pressure, getting that movie together, all the media attention."

George nodded. "It's true. He's been handling everything so expertly, making me feel a total fool about the past – but it couldn't have been easy." He faced her directly for the first time. "So what is the story about my brother?"

"Not good, if I understood Martin correctly. They think he hasn't got long." She hesitated. "Cancer."

"I need to think about this..."

"I understand. Martin told me there was some bad blood."

"So he did see you then?" he said, stupidly. Of course the boy had.

"Twice – three times now. The second time, he just dropped off a picture of his mother he'd borrowed to copy. I also gave him his baby photos. Didn't he tell you?"

"No. But he did say he wanted to see you."

Geraldine nodded. "He very much wanted to know more about his mother. I was surprised he wasn't angry with me. I still don't feel what we did was any more wrong than any other kind of break-up, but they are seldom easy for a small child."

"Yes. He was very understanding – taught me a few things about how I should have handled it. But I must say, he was pleasantly surprised when I told him how I stood up for her against my family."

"And church."

"He told you that too?"

"Look, George. I don't know if we can ever be friends, but Martin and I have a kind of understanding – almost a bond, somehow. He was in desperate shape when he landed up on my doorstep today. He must have headed for me almost by instinct. Whether we like it or not, I'm part of his life. I won't show up here again if it makes you uncomfortable – but this seems the right time to mention something else."

George leant forward in his chair.

"He asked me if I could be at his wedding –" she held up a hand, stopping his response – "but *only* if you would be happy to have me there."

"If that's what he wants..."

"No, George. Be sure what *you* want. This boy is still a bit fragile about being fibbed to by people close to him. If I'm there and you look uneasy, it will be a terrible situation. I won't have it, and you shouldn't contemplate it. Tell it to him straight."

"I see. I have a lot to think about."

As she was shown to the door, Geraldine stopped. "Oh, and one more thing.

"When Lucy died, you know, I thought I was a very strong person, but I ended up in much the state Martin was in today. I was in therapy for six months, and it was worth it."

"Therapy?"

"Don't argue, George. I think the two of you should try a couple of sessions. My therapist is really good. I'll give you his details. You won't regret it."

"So many things to think about..."

"No navel contemplation. I learnt the hard way that you don't solve a problem that way. I'll go back to check on the lad, and give you a call from there." She stopped at the foot of the stairs, short of her car. "Do you have a friend you can talk to?"

"Yes." He glanced next door. Muriel's house looked occupied – car there, lights on. "Yes, I do."

Back at Martin's, the place looked dark from outside. Geraldine gently tapped on the door, and half a minute later, Angie opened it, shortly after a light turned on. She looked dazed, a week short on sleep.

"How is he?"

"Sleeping."

"Did I wake you up?"

"No. Couldn't sleep."

"I talked to his dad. Passed on the message, so that's out of the way."

"Who is it?" Martin's voice sounded strong.

"Geraldine," replied Angie.

Martin padded to the door, suddenly remembering to check if he was wearing anything. He had shorts on.

"Geraldine, I'm so sorry – I was such a wreck. You were so good, and Julie – I don't know how long she was here – I can't imagine what it looked like to all of you. Is my dad. . . ?"

"Martin, he's your *dad*. Until he goes senile, you don't run his life. I'm sure he's going through some difficult choices right now, but you don't have to fix everything. Take a day off. Smell some flowers. Go for a long walk with Angie, whatever it takes to unwind. In the meantime, I told him I'd let him know how you were when I got here." She looked around. "Where's your phone?"

* * * * *

He is talking on the phone to British Antarctic Survey people, going through images and footage he could use.

James McLaughlan at Bird Island tells about all the precautions against importing rats, how the fur seal numbers are recovering since hunting was stopped.

"What can we learn about climate change from the island?"

"Since we don't have permanent snow and ice here, we are less affected than Antarctica or Greenland by climate change, so we can use populations here as a baseline to compare populations where ice is being lost."

He is talking to someone at Halley V Base. George Burns. "So, Dr Burns, how long have you been at this Antarctic research station?"

"A few weeks. We have a wintering staff who keep the place going, but we can't do much science over winter – the weather's too extreme. Right now, it's a sultry minus twenty nine celsius."

"How old is this station?"

"We've been active at this site since 1956, but this is the fifth incarnation of the buildings. Previous ones were eventually crushed by snow build-up. This one is designed to move."

"So you have many years of history at that site. What are the trends?"

"There's a limit to what you can measure on the ground. Of course we know that massive ice shelves have broken up."

"Larsen B."

"The famous one. But the best data we're getting is from satellite measures – very fine gravitational measures of the mass of the ice, which clearly show that the ice is thinning on the edges."

Pictures of emperor penguins, stock footage, discussion of their incredible adaptation to the Antarctic – and how radical changes to the ice sheets could be the end of them.

Editing. More editing. Julie drifting in and out of focus, Angie, always there – chasing up after interviewees who wanted the right to rebuttals without deadlines – more editing, burning backups. . .

Therapy sessions. His dad. Muriel, increasingly a presence.

One day after a heavy editing session, he found a couple of books on the coffee table. Richard Dawkins: *The God Delusion*; John Shelby Spong: *Jesus for the Non-Religious*. "Dad. What's this? I thought you gave up on churches a long time ago."

"Interesting stuff – Muriel put me onto it."

Martin nodded. "Good to see the two of you getting on so well. The movie is really close to done now. We should all have a party when..."

He was taking a pile of DVDs to Prentice's office – his preferred edit and some variants, allowing for differences in timing for the various contracts Prentice had negotiated. Angie was with him.

There were effusive greetings, Prentice this time pointing out his various trophies and memorabilia, the stories behind them. Hand shakes, more signing of documents, impossible-looking dollar amounts...

They were all together at his dad's home, Angie, his dad, Muriel, Geraldine, Julie and family.

The titles rolled over a scene of emperor penguins.

Martin's face. "I'm Martin Truscott.

"There has been considerable debate about climate change, probably too much.

"On the one hand, IPCC-aligned scientists say the science is settled. On the other, a group calling themselves 'sceptics' claim it's anything but settled – that the IPCC position is alarmist.

"Where does the truth lie? That's the question I set out to explore, and I invite you to join me."

Scene of Schoor fleeing from the TV studio in Boston.

"One thing I want to make clear: the confrontation I had with Professor Schoor is a side show. The important thing is, as always, the science. So let us focus on that."

Blunt on screen, being asked about burning fossil fuels as

if there's no tomorrow: "No tomorrow? That could turn out to be prophetic."

Martin back on screen: "So let's look at the science: Dr Blunt has been accused of being alarmist."

Interviews, inter-cut with Martin summarizing, as the case built up.

Finally: "There we have it. The case as well as I could put it, based on what we know now.

"Is Blunt's position of multi-metre sea level rise by the turn of the century 'alarmist'? We can't be absolutely sure. But what we do know is that sea level is rising faster than the IPCC predicts.

"Is the contrarian position that there is no evidence of human-induced warming correct? I couldn't find the strong evidence they claimed.

"Ultimately, the bottom line is this." He paused. "Do we believe scientists who have built up a mass of evidence that they can clearly demonstrate, or contrarians whose evidence is hard to find, and even harder to support?

"But don't take my word for it – I have assembled all the evidence I could find on a web site." Details flashed up on the screen for long enough to be written down. "I am not an expert, but I have spoken to many experts, and learnt a lot – much more than I expected – making this movie. That's what science is ultimately about: learning more than we expected."

*   *   *   *   *

It was a bright, early summers day, and the Greens were up reasonably early, almost into the time zone. George Truscott had them awake and chatting on the front deck, when Martin and Angie showed up.

Martin took the lead. "Hi, Dad. Bob. Melissa." He and Angie took turns hugging her parents.

George said, "My family was never all that big on hugging and kissing."

"Your son is." Angie flashed him her most impish grin.

"Oh, well, can't beat them..."

He put both arms on her shoulders, closed his eyes, then pulled her close. She held on when he tried to let go. "There, there. That didn't hurt a bit," she said reassuringly.

He opened his eyes, and returned her smile.

Martin gave him a gentle hug. "Dad, it is kind of odd when you aren't used to it, but I got over it pretty fast."

"You know, George," Bob opined, "you did a great job with Martin. We've been very lucky in many ways – Melissa and I hit it off from the start, Angie has always been ahead of everyone including us, but look at him, all over national – no *international* – TV, shaking up climate change complacency..."

"Well, I'm not about to publish the Truscott parenting manual."

Martin had never seen the old man so at a loss for words. "Dad, you must admit, it was pretty hard – you had no family support, your kid was a terrible brat who always did the opposite to what he was told..."

"Hmm, yes. Reverse psychology..."

"Dad, you didn't!" Martin was horrified at the thought that he might have been manipulated by yet another person close to him – even if his dad hadn't seemed that close at the time.

"Well, you were dead set on that game thing. Even though you didn't really try at school, you were acing the maths courses. I really thought you would be a great engineer, but you weren't interested. So I went to a few unis, talked to a few people in the industry. Turned out you needed pretty much an engineering background to get where you wanted to be. So I decided not to fight it – too hard. Just made sure that you picked up some hints about the right subjects to take, and kept on about how good it would be if you switched to engineering. Maybe you refused to do engineering, but you ended up mostly doing engineering subjects. You could've picked up an engineering degree in maybe eighteen months, with credits for previous study."

George smiled. "Sorry, I don't have that wicked grin thing Angie taught you, but I like to feel I did give you a few pushes in the right direction.'

"You *manipulated* me? It didn't feel like it a whole lot at the time."

"No, boy. It was mostly you. You surprised me sometimes, even if I was plenty worried about you. I remember one night..."

*   *   *   *   *

George Truscott was heading for bed.

Martin was at his computer again. It was getting late. His dad stopped at his door. "Not playing damn games again? You can't get a job doing that."

"You don't know what the hell I'm doing."

"What I do know is, I'm your dad, and I don't want you to screw up your life."

"I know what I'm doing."

"Well, let me tell you one thing you don't know. In the real world, you need to deal with people. Sometimes people you don't like. That's life. Maybe you don't like your dad too much. Maybe one day you're working for someone you don't like too much, but it's the job you want. What do you do?"

"So who are you, then? The pointy-haired manager?" Martin sneered.

"The pointy-haired manager?" For a moment, the boy had him.

Then Martin clarified: "Dilbert..." The ultimate insult to an engineer: to be on the wrong side of the Dilbert menagerie.

"Oh." From the look on the boy's face, he could see how he'd stung his dad's engineering pride. "Just don't stay up all night playing games."

"Go to bed dad. It's past your bed time. You're too old to be up this late."

He gave up and went to bed.

He woke in the early hours of the morning, needing to go to the bathroom. He saw a light under his bedroom door. He opened the door, and sure enough, Martin's light was still on. "Stupid," he muttered, "doesn't he have something first thing tomorrow?"

If the boy was playing a game, it was a very quiet one. George padded over, to find his son asleep at his desk, his face in a calculus text. Next to him was a tutorial sheet, with spaces for answers. The tutorial session was tomorrow (technically later that morning, he realized), 8 am.

Most of the answers were filled in. From what he could remember of his engineering calculus, they looked good. He shook his head. Gently, he pried his son upright, and maneuvered him to bed, took his shoes off and pulled a sheet over him. The boy murmured something that sounded like "partial derivative".

He backed out of his son's room, and turned the light off. "I never did a tute ahead of time in my life. Well, not so close to the whole thing, for sure."

*   *   *   *   *

"Dad, did that happen a lot – you spotting me doing stuff you didn't expect?"

"Of course it bloody did. You don't mean to say you looked up your course results on my computer, then left them visible by mistake? Or have you forgotten the times you did it... that mischievous grin again – huh! Look at you! You remember the 105% one I bet."

"Well, I thought getting 105% was pretty cool, and it would get you off my back."

"105%? How did you do that?" Angela demanded.

"Lecturer was tired of being nagged for extensions so he offered a bonus for early hand-in. It was a challenge. I hit every point on the mark scheme, and handed in early. He had to give me 105%."

"Cool. What was the course?"

"I forget. Algorithms or something."

"Before you met Julie?"

"Uh, yes, Angie. Not everyone in my life makes me perform academically. Just...my dad, as I've now discovered, and – I very much hope – you."

"Hang on." George Truscott went off to his study.

He came back with a dog-eared piece of A4 paper, faded with the years. It was a printout of Martin's results from the course web page with a detailed breakdown and the 105% total at the end.

Martin took it from him. "You kept that all these years?"

"You wanted me to see it. I hoped there would be a day when I could say, 'well done', and have you take it the right way."

Martin gave his dad a bear hug, pleased to feel his old man less stiff this time.

Angie grabbed the piece of paper while Martin was distracted. "I never ever got 105%. I kind of felt it was OK to slack off when I did everything needed to get to 100%."

"Well. I didn't do this every time." Martin went a bit pink.

"Better get a towel." Everyone looked at Angie in bewilderment.

Towel? What the...Martin unravelled the logic...Towel. Shower? Doesn't make sense...Swimming. The pool. Berkeley.

"Modesty." As he said this, they collapsed into each other's arms laughing.

It took some explaining, less to Angie's parents, as they only needed to be reminded. "Sorry," Angie said to George, "you had to be there."

*　　*　　*　　*　　*

Getting a whole family in for a single shot was always a challenge, what with little kids wandering off, and older people who hadn't seen each other since the last millennium getting into conversation and losing focus. How much easier would it be for a professional photographer? But he was pleased to be doing it – amateur or no.

Eventually, it was all about right. "Mabel, deary, you're a bit short for that position – stand on the step behind."

Last adjustment out of the way, Henry Truscott rushed over to stand next to his brother before the flash went off.

"Just wait. . . one more."

He fiddled with the camera again. "Geraldine, dearie, you're sticking out on the edge. Snuggle into Geraldine a bit closer, if she doesn't mind."

Another flash, then individual pictures – Marty and Angie, until they looked too tired to stand, George with his surviving brother (rushing between his camera and his pose) and great aunt. Julie with her family and Marty. Marty and Angie with his maternal grandmother. Julie's parents. And all variations involving Marty and Angie separate with the other combinations and together.

Not such a big party as these things went, but nothing could be left out.

Finally Henry packed his gear, and shook Martin's hand, a firm handshake on both sides.

"Good to meet you finally, boy."

"Good that we have all that old crap behind us," Martin said, solemnly. The look behind his eyes said, yes – that is really behind us now.

20   Back on the Ground

MARTIN PARKED THE CAR near the cellar door. "This is one good reason to move from Queensland to New South Wales. They try to make wine in various spots, but nothing there compares with Hunter Valley – wrong climate, really. Though dad tells me this area wasn't so flash when it started and one or two Queensland places are half decent – despite the odds."

Angie smiled. "When we go to visit family in California, we should go to Napa Valley."

They didn't rush to go into the building. Martin looked around. "Why don't we walk around a bit? It's so nice to be out of the city."

"When last did we do something like this?" she asked, as they walked along the rustic road.

"Stanford?"

"Probably," she added.

They walked silently for a bit, the solitude occasionally interrupted by cars going to one or another winery.

"It's funny," he eventually said.

"What?"

"When I mentioned Stanford, it got me thinking, not many people start a PhD to slow down a bit."

"You really did need to slow down – that last six weeks or so up to finishing the movie was scarily intense."

"I know. After that thing with dad's aunt phoning, it was as if, I don't know – a balloon had burst – all the pressure came out at once."

"You were fantastic on those radio talk shows and TV panels. Totally demolished the bogus economics arguments."

"Maybe. I hardly remember. It really isn't my debate anyway – everyone should be trying to understand the science – or accepting the mainstream position."

"Which didn't include massive sea level rise until you forced everyone to think about it."

"Did I *really* do that?"

"Where's a towel when we need one most?"

He hugged her, oblivious to passing cars.

"Of course you did," she went on. "A Chinese government TV channel wouldn't have bought your movie if they weren't about to change course."

"But still so much to do. At least my modelling is starting to come together. Those NASA people, the British bunch – they're all really good. This stuff is way harder than that game stuff I used to do."

They walked on silently for a bit.

"I don't understand too much detail of your bio stuff, just some of the compute issues – does it help you, talking to me about it?"

"Oh yes. My advisor keeps telling you have to have an elevator pitch – something you can tell someone before the doors open again that summarizes the main point – otherwise you are drifting and have no focus."

He looked disappointed. "What's the long face for?" she asked.

"I want you to be forced to keep talking to me forever."

She laughed. "Don't worry, I'm not going anywhere."

He looked troubled, and stopped walking. "When I ran away from you that time – the phone call – we've never talked about that." He gazed into the distance. "I'm so sorry." He turned back to her. "That must have been horrible for you. Why did you bring in Julie?"

"I didn't really know anyone else there except your dad, and I thought Julie would have a better idea of places you might hang out."

"From what little I remember of that I don't think any of it was 'hanging out'."

"Never mind. As far as I can tell, the therapy worked – and getting that movie out put that all in the past, somehow too. What did Geraldine keep telling us? No navel contemplation? I'm just so pleased to have my special guy back."

"What if I hadn't snapped out of it?"

"I would still be working on it."

They walked on. At a view site, they stopped.

"Do you remember that conversation we had, about one of us staying home to make babies?"

She thought for a bit. "Oh yes, you said you'd do that. I sometimes forget you aren't a biologist. Should I explain? Or maybe develop a computer model so it's a bit clearer for you?"

He laughed. "No. I'm just thinking though. You are dead

set on being an academic. I'm only doing this because I want to solve this one problem. Maybe I should be the one to stay at home parenting..."

"That sounds dangerously like navel contemplation – we have PhDs to finish."

"Forfeit?"

"I graciously accept. I'll let you decide."

"I think I'll buy you some really nice wine."

"Not much of a forfeit – you get your share."

"Not much of a crime."

"True. Let's head back and do some shopping."

After walking a bit, something else popped into his head. "Your band." She looked puzzled. "I picked up somewhere that you tried to break into music at some point in the past."

"Why bring *that* up now?"

"Just curious. I heard this thing a while back, and never got around to following up." She gave him a look. "You know, you and your buddies. Didn't work because you were all tone deaf."

"Mom!" Ange shouted loud enough to be heard in California. "She told you didn't she?" His face was a rigid mask. "Oh well, since you ask. We had this great plan of being the first purely internet band. Music free to download, the money would be in gigs and concerts. I was like twelve, and thought I knew everything. We saved up allowances, did extra chores, and so on. Bought video cameras, instruments, paid for our own web site – we had the whole thing nailed."

"And no one noticed?" Martin offered the most common reason for failed ecommerce.

"Worse. We had a review in *Rolling Stone*. A *very* short

review. 'If this is the first purely internet band, we sure hope it's the last.' Had to pull the web site down, it was getting so many hits – but no invitations to do gigs."

About halfway back, she looked contemplative. "Something I've been meaning to ask. That brother of your dad's, the photographer. Didn't he sound a bit gay to you, calling everyone 'deary'?"

"Dunno. I don't pay attention to things like that." He shook his head. "Poor old Henry. If that's true, he must have been really conflicted over how they all reacted to my mum – if he isn't so deep in the closet, he doesn't know where he is." A few paces later: "Speaking of which, I wonder how dad is taking Canberra?"

Angie worked up an especially impish grin. "More to the point, how is Canberra taking dad? A conservative engineer backing the gay rights cause – it's probably already a big enough shock for him to have gone green."

As they continued to walk back, he started placing his feet very deliberately. After a few paces like that she pointed at his feet and asked, "What's that for?"

"So nice to have my feet back on the ground."

21   Tomorrow

THE LITTLE BOY ran down to the end of the garden.

There were tall shrubs between him and the house, making it feel like his very own space.

He found a shady spot under the old mango tree, and sat down, playing with the shifting shadows.

A small bird fluttered down, its cheeky long tail sticking out. It hopped around. He watched it. Another bird landed, just like the first. The pair of them hopped closer, inquisitively.

An elderly voice floated down distantly from the house. "Tea's ready."

He watched the hopping birds, immobile.

They hopped closer.

He put out a hand as if to touch, but stopped well short of the nearest bird.

Both birds examined him quizzically. He smiled shyly and left his hand extended.

The birds hopped closer.

Afterword

This is a work of fiction.

Many of the places and institutions mentioned are real. NASA's Goddard Institute for Space Studies is real. There are genuinely people there who contribute to `realclimate.org` (which is also real), and who are leaders in pressing the case for urgent action for example against the risk of rapid ice cap melts. The actual characters of course are made up. I do not personally know anyone at NASA other than from reading `realclimate.org` (and the occasional chance meeting in an academic setting on other unrelated matters).

There really is a hostel called the Blue Parrot, but it's in Sydney, Australia, not Boston, MA. I've been to the real one, and only chose to recycle the name because I liked it.

The Union Oyster House is real and well worth a visit if you're in Boston.

Harvard and MIT of course are real. Though I made up the people from those institutions, look up participants in the climate change debate to see for yourself whether I properly positioned these otherwise great institutions fairly in the big picture.

The *Wall Street Journal* has been just as sloppy in its opinion pages as reported here. Feel free to check – though they hide behind pay-to-view web access to reduce the embarrassment.

The Australian universities mentioned are all real, though none of the Australian personalities mentioned are. There is a British Antarctic Survey, which really is a major player in climate research. The people again are made up.

The ABC really is Australia's approximation to the BBC – and a good one, for a small country. SBS is a real channel as well: also government funded, but with a commercial slant.

The IPCC, or Intergovernmental Panel on Climate Change, is real of course.

There really was a climate change movie (made for TV) with "Swindle" in its name, and Martin Durkin really did make it. It really was poorly made. It started from the premise that the mainstream climate scientists were a pack of liars, and presented highly selected "facts" to make the case. He was pulled apart in Australia by one of the ABC's top political interviewers, Tony Jones, much to the chagrin of the denial camp. "Why didn't they do that to Al Gore?" Well, Al Gore didn't start from the position that everyone with contrary views was lying, and his science largely stands up to scrutiny. He didn't have to change a whole lot of details after the initial release of his movie resulting from exposure of embarrassing errors.

Since there are people out there disputing the science, whether for malevolent motives, or because they have staked out a position and don't want to admit they are wrong – I have yet to find one who is really a Galileo, right in the face of a determined Inquisition – I have tried to be accurate with the science to the extent that it's known. The potential for a major collapse of the West Antarctic Ice Sheet is not clearly understood, and I have deliberately avoided extrapolation on that point – only proposed that this possibility be taken more seriously. However, the worrying thing is that the science is becoming *less* certain, and the certainty that's eroding is mostly on the side of "it won't happen". Some examples aren't mentioned in the narrative: movement of parts of the ice is much faster than has previously been predicted, based on glacial studies. The IPCC predicted that increased precipitation would mostly compensate for increased melting. However, what has actually happened is the interior of the ice sheet has thickened, while the edges are coming away faster than expected. This is especially bad news for WAIS, because about half of it is below sea level, up to 2km below – as reported in the book. Any tendency for the ice to break up could rapidly accelerate if the edges that are over water break away, because ice floats. While temperature change is more or less in the middle of the IPPC "consensus" predictions, sea level change is increasing *faster* than any model the IPCC accepts. The IPCC has *not* reduced its prediction of sea level change as some "sceptics" claim. The apparent reduction between their 2001 and 2007 numbers does mainly result from removing figures for ice cap melts, as reported in the book. So the highest sea level rise in the IPCC model now *excludes* the one measure we know is growing faster than expected. Not cause for optimism.

Is the first sentence of the previous paragraph fair? Could some be disputing the science because they genuinely aren't convinced? This is what our protagonist set out to find out. He failed to find these people or at least to be convinced by them. Did he try hard enough? The number of genuine climate scientists with real doubts is small, but some do exist. However, the ones who get the press are generally not credible: they are not actively researching the area. They have diverse motivations: for example, some have a gut feel that anything that is pro-environment must be wrong, based on their experience in the distant past, when environmentalism was the province of emotional tree huggers. Of course, some also are industry shills. You need to understand a fair amount of the science to tell these and other cases apart – but you don't need to be an expert.

A reader of the first edition asked me why I didn't touch on the economic arguments: why we shouldn't focus all our efforts on third world poverty, instead of addressing global warming. This argument doesn't stand up to scrutiny, and shouldn't have been given the space it was given in *Swindle*. First, there's nothing to stop anyone from working to alleviate poverty, and it amazes me that a whole lot of Albert Schweitzers emerge from the woodwork with arguments like this when corporate interests are threatened. Where were they before global warming was an issue? Second, it is by no means clear that climate change mitigation will disadvantage poorer countries. Superior, cheaper renewable solutions will benefit countries without energy grids significantly more than countries which already have a heavy investment in existing technology. Finally, the poor will be hit hardest by wild swings in climate whether in the form of crop failures, or sea level rise. They have less to fall back on, and fewer resources to fight back. I could have worked all this into the narrative, but it seemed pointless, because the arguments are so easily dismissed, and it's hard to imagine that anyone but a hard-core denialist really believes them.

There are numerous other matters I could have touched on but didn't in the interests of maintaining a narrative rather writing a polemic. For example, every time we get a cooler year than the previous, we get told that this means warming has ended (a generalisation of the "global warming ended in 1998" claim). Of course, there is natural variability in the climate: you get warm and cool years. The warming trend is superimposed on this natural variability. To see the long term trend, you need to use the same trick as stock analysts use: look at a moving average. For example, a 5-year moving average is calculated by starting at a year, and averaging the last 5 years. Plot a point. Now, move on to the next year, and average over the last 5 years to get the next point. If you do this with the temperature trend, even if there is the odd warmer or cooler year, the trend is consistently up.

Explore the matter yourself. Start at realclimate.org. Read the "sceptic" literature. Read the accepted science. It's a pretty big job. You also need to do considerable intelligent filtering because anyone can write whatever they like in a blog; try to get to original sources where you can. Fortunately, some of the big players like NASA publish all their papers on their web site (far too many other papers, unfortunately, require expensive subscriptions to read). Someone has to be wrong. If it's the mainstream, the worst that will happen is that we will pay a bit extra for a cleaner planet. If it's the "sceptics", hundreds of millions of lives will be disrupted or lost. Unfortunately, we don't have two different planets in which to try the alternative experiments. Small stakes? No wonder some are playing hardball.

To find the best material do not focus primarily on blogs and articles you can find with ordinary web searches. Anyone can publish anything on the net. Google Scholar on the whole tends to find better articles:

http://scholar.google.com/

Finally I would like to thank those who have helped with this project. The School of Information Technology and Electrical Engineering at the University of Queensland was good enough to host me as a Visiting Fellow while I was working on this book. John Shield offered suggestions to improve an early draft. Evan Jones offered a few hints. My wife Fiona Semple read through everything in detail and found a fair number of corrections. Don Eldridge was good enough to write a detailed critique of the first edition. Nhanhla Mabaso found a few more errors, corrected in the second printing of the second edition. After all this, Lyn Aberdeen found a few more typos and Richard Rodseth found a further list of corrections. Everything in here is ultimately my judgment call: I didn't act on suggestions if I didn't like them or agree with them but all these people collectively, nonetheless, contributed to the book as you find it now.

This book has a web site where you can read reviews and post your own comments:

http://groups.google.com.au/group/no-tomorrow

www.ingramcontent.com/pod-product-compliance
Lightning Source LLC
Chambersburg PA
CBHW071514260626
47170CB00002B/359